Fault-Tolerant Design

Elena Dubrova

Fault-Tolerant Design

Elena Dubrova
KTH Royal Institute of Technology
Krista
Sweden

ISBN 978-1-4939-0240-8 ISBN 978-1-4614-2113-9 (eBook)
DOI 10.1007/978-1-4614-2113-9
Springer New York Heidelberg Dordrecht London

Softcover re-print of the Hardcover 1st edition 2013

Printed on acid-free paper

Springer is part of Springer Science+Business Media (www.springer.com)

To my mother

Preface

This textbook serves as an introduction to fault tolerance, intended for upper division undergraduate students, graduate-level students, and practicing engineers in need of an overview of the field. Readers will develop skills in modeling and evaluating fault-tolerant architectures in terms of reliability, availability, and safety. They will gain a thorough understanding of fault-tolerant computing, including both the theory of how to achieve fault tolerance through hardware, software, information, and time redundancy and the practical knowledge of designing fault-tolerant hardware and software systems.

The book contains eight chapters covering the following topics. Chapter 1 is an introduction, discussing the importance of fault tolerance in developing a dependable system. Chapter 2 describes three fundamental characteristics of dependability: attributes, impairment, and means. Chapter 3 introduces dependability evaluation techniques and dependability models such as reliability block diagrams and Markov chains. Chapter 4 presents commonly used approaches for the design of fault-tolerant hardware systems, such as triple modular redundancy, standby redundancy, and self-purging redundancy and evaluates their effect on system dependability. Chapter 5 shows how fault tolerance can be achieved by means of coding. It covers many important families of codes, including parity, linear, cyclic, unordered, and arithmetic codes. Chapter 6 presents time redundancy techniques which can be used for detecting and correcting transient and permanent faults. Chapter 7 describes the main approaches for the design of fault-tolerant software systems, including checkpoint and restart, recovery blocks, N-version programming, and N self-checking programming. Chapter 8 concludes the book.

The content is designed to be highly accessible, including numerous examples and problems to reinforce the material learned. Solutions to problems and PowerPoint slides are available from the author upon request.

Stockholm, Sweden, December 2012 Elena Dubrova

Acknowledgments

This book has evolved from the lecture notes of the course "Design of Fault-Tolerant Systems" that I have taught at the Royal Institute of Technology (KTH), since 2000. Throughout the years, many students and colleagues have helped me polish the text. I am grateful to all who gave me feedback. In particular, I would like to thank Nan Li, Shohreh Sharif Mansouri, Nasim Farahini, Bayron Navas, Jonian Grazhdani, Xavier Lowagie, Pieter Nuyts, Henrik Kirkeby, Chen Fu, Kareem Refaat, Sergej Koziner, Julia Kuznetsova, Zhonghai Lu, and Roman Morawek for finding multiple mistakes and suggesting valuable improvements in the manuscript. I am grateful to Hannu Tenhunen who inspired me to teach this course and constantly supported me during my academic career. I am also indebted to Charles Glaser from Springer for his encouragement and assistance in publishing the book and to Sandra Brunsberg for her very careful proofreading of the final draft. Special thanks to a friend of mine who advised me to use the MetaPost tool for drawing pictures. Indeed, MetaPost gives perfect results.

Finally, I am grateful to the Swedish Foundation for International Cooperation in Research and Higher Education (STINT), for the scholarship KU2002-4044 which supported my trip to the University of New South Wales, Sydney, Australia, where the first draft of this book was written during October–December 2002.

Contents

Acronyms

ADSL	Asymmetric Digital Subscriber Line
ALU	Arithmetic Logic Unit
ASIC	Application Specific Integrated Circuits
ATM	Automated Teller Machine
ATSC	Advanced Television Systems Committee
CPU	Central Processing Unit
CRC	Cyclic Redundancy Check
DRAM	Dynamic Random Access Memory
DSL	Digital Subscriber Line
DVB	Digital Video Broadcasting
DVD	Digital Versatile Disc
FD	Fault Detection
FIT	Failures in Time
FPGA	Field-Programmable Gate Array
IC	Integrated Circuit
I/O	Input/Output
LFSR	Linear Feedback Shift Register
MTTB	Mean Time between Failures
MTTF	Mean Time to Failure
MTTR	Mean Time to Repair
NMR	N-Modular Redundancy
PC	Parity Checker
PG	Parity Generator
POSTNET	Postal Numeric Encoding Technique
RAID	Redundant Array of Independent Disks
RAM	Random Access Memory
SoC	System-on-Chip
TMR	Triple Modular Redundancy

Chapter 1
Introduction

"If anything can go wrong, it will."
Murphy's law

In this chapter, we formally define fault tolerance and discuss its importance for designing a dependable system. We show the relation between fault tolerance and redundancy and consider different types of redundancy. We briefly cover the history of fault-tolerant computing and describe its main application areas.

1.1 Definition of Fault Tolerance

Fault tolerance is the ability of a system to continue performing its intended functions in presence of faults [7]. In a broad sense, fault tolerance is associated with reliability, with successful operation, and with the absence of breakdowns. A fault-tolerant system should be able to handle faults in individual hardware or software components, power failures, or other kinds of unexpected problems and still meet its specification.

Fault tolerance is necessary because it is practically impossible to build a perfect system. The fundamental problem is that, as the complexity of a system grows, its reliability drastically decreases, unless compensatory measures are taken. For example, if the reliability of individual components is 99.99 %, then the reliability of a system consisting of 100 non-redundant components is 99.01 %, whereas the reliability of a system consisting of 10,000 non-redundant components is just 36.79 %. Such a low reliability is unacceptable in most applications. If a 99 % reliability is required for a 10,000-component system, individual components with a reliability of at least 99.999 % should be used, implying a sharp increase in cost.

Another problem is that, although designers do their best to have all the hardware defects and software bugs cleaned out of the system before it goes on the market, history shows that such a goal is not attainable [3]. It is inevitable that some unexpected environmental factor is not taken into account, or some potential user mistakes are not foreseen. Thus, even in the unlikely case that a system is designed and

E. Dubrova, *Fault-Tolerant Design*, DOI: 10.1007/978-1-4614-2113-9_1,

implemented perfectly, faults are likely to be caused by situations outside the control of the designers.

A system is said to *fail* if it has ceased to perform its intended functions. *System* is used in this book in the generic sense of a group of independent but interrelated elements comprising a unified whole. Therefore, the techniques presented are applicable to a variety of products, devices, and subsystems. *Failure* can be a total cessation of function, or a performance of some function in a subnormal quality or quantity, like deterioration or instability of operation. The aim of fault-tolerant design is to minimize the probability of failures, whether those failures simply annoy the users or result in lost fortunes, human injury, or environmental disaster.

1.2 Fault Tolerance and Redundancy

There are various approaches to achieving fault tolerance. Common to all these approaches is a certain amount of redundancy. For our purposes, *redundancy* is the provision of functional capabilities that would be unnecessary in a fault-free environment [8]. This can be a replicated hardware component, an additional check bit attached to a string of digital data, or a few lines of program code verifying the correctness of the program's results. The idea of incorporating redundancy in order to improve the reliability of a system was pioneered by John von Neumann in the 1950s in his work "Probabilistic logic and synthesis of reliable organisms from unreliable components" [12].

Two kinds of redundancy are possible [1]: space redundancy and time redundancy. *Space redundancy* provides additional components, functions, or data items that are unnecessary for fault-free operation. Space redundancy is further classified into hardware, software, and information redundancy, depending on the type of redundant resources added to the system. In *time redundancy* the computation or data transmission is repeated and the result is compared to a stored copy of the previous result.

1.3 Applications of Fault Tolerance

Originally, fault tolerance techniques were used to cope with physical defects in individual hardware components. Designers of early computing systems triplicated basic elements and used majority voting to mask the effect of faults [11].

After World War II, the interest in fault tolerance increased considerably. One reason for this was the reliability problems with electronic equipment which occurred during the War [2]. It became clear that future military technology would rely heavily on complex electronics. Another catalyzing factor was the "space race" between the United States and the Soviet Union which followed the launching of the first Earth satellite Sputnik in 1957 [2]. As a result, defense departments and space agencies in many countries had initiated projects related to fault tolerance.

As semiconductor technology progressed, hardware components became intrinsically more reliable and the need for tolerating component defects diminished for general purpose applications. Nevertheless, fault tolerance remained an essential attribute for systems used in safety-, mission-, and business-critical applications. *Safety-critical* applications are those where loss of life or environmental disaster must be avoided. Examples are nuclear power plant control systems, computer-controlled radiation therapy machines, heart pacemakers, and flight control systems. *Mission-critical* applications stress mission completion, as in the case of a spacecraft or a satellite. *Business-critical* applications are those in which keeping a business operating continuously is an issue. Examples are bank and stock exchange automated trading systems, Web servers, and e-commerce.

During the mid-1990s, the interest in fault tolerance resurged considerably. On the one hand, noise margins were reduced to a critical level as the semiconductor manufacturing process size continuously shrinks, the power supply voltage is lowered, and operating speed increases. This made Integrated Circuits (ICs) highly sensitive to transient faults caused by crosstalk and environmental upsets such as atmospheric neutrons and alpha particles [10, 14]. It became mandatory to design ICs that are tolerant to these faults in order to maintain an acceptable level of reliability [6].

On the other hand, the rapid development of real-time computing applications that started around the mid-1990s, especially the demand for *software-embedded* intelligent devices, made software fault tolerance a pressing issue [9]. Software systems offer a compact design and a rich functionality at a competitive cost. Instead of implementing a given functionality in hardware, a set of instructions accomplishing the desired tasks are written and loaded into a processor. If changes in the functionality are required, the instructions can be modified instead of building a different physical device.

Software eliminates many of the physical constraints of hardware; for example, it does not suffer from random fabrication defects and does not wear out. However, an inevitable related problem is that the design of a system is performed by someone who is not an expert in that system. For example, the automotive brake expert decides how the brakes should work and then provides the information to a software engineer, who writes a program which performs the desired functionality. This extra communication step between the engineer and the software developer is a source of many faults in software today [3].

In recent years, the interest in fault tolerance has been further boosted by the ongoing shift from the traditional desk-top information processing paradigm, in which a single user engages a single device for a specialized purpose, to *ubiquitous computing* in which many small, inexpensive networked processing devices are engaged simultaneously and distributed at all scales throughout everyday life [13]. Industry foresees that over 50 billion people and devices will be connected to mobile broadband by 2020 [5]. As our society turns into a "networked society", it becomes increasingly important to guarantee the dependability of all services and players involved in the network. It is likely that new, nontraditional approaches to achieving fault tolerance will be needed [4].

References

1. Avižienis, A.: Fault-tolerant systems. IEEE Trans. Comput. **25**(12), 1304–1312 (1976)
2. Avižienis, A.: Toward systematic design of fault-tolerant systems. Computer **30**(4), 51–58 (1997)
3. Brooks, F.P.: No silver bullet: essence and accidents of software engineering. IEEE Comput **20**(4), 10–19 (1987)
4. Dubrova, E.: Self-organization for fault-tolerance. In: Hummel, K.A., Sterbenz, J. (eds.) Proceedings of the 3rd International Workshop on Self-Organizing Systems, Lecture Notes in Computer Science, vol. 5243, pp. 145–156. Springer, Berlin/Heidelberg (2008)
5. Ericsson: More that 50 billions connected devices (2012). www.ericsson.com/res/docs/whitepapers/wp-50-billions.pdf
6. ITRS: International technology roadmap for semiconductors (2011). http://www.itrs.net/
7. Johnson, B.: Fault-tolerant microprocessor-based systems. IEEE Micro **4**(6), 6–21 (1984)
8. Laprie, J.C.: Dependable computing and fault tolerance: concepts and terminology. In: Proceedings of 15th International Symposium on Fault-Tolerant Computing (FTSC-15), pp. 2–11 (1985)
9. Lyu, M.R.: Introduction. In: M.R. Lyu (ed.) Handbook of Software Reliability, pp. 3–25. McGraw-Hill, New York (1996)
10. Meindl, J.D., Chen, Q., Davis, J.A.: Limits on silicon nanoelectronics for terascale integration. Science **293**, 2044–2049 (2001)
11. Moore, E., Shannon, C.: Reliable circuits using less reliable relays. J. Frankl. Inst. **262**(3), 191–208 (1956)
12. von Neumann, J.: Probabilistic logics and synthesis of reliable organisms from unreliable components. In: Shannon, C., McCarthy, J. (eds.) Automata Studies, pp. 43–98. Princeton University Press, Princeton (1956)
13. Weiser, M.: Some computer science problems in ubiquitous computing. Commun. ACM **36**(7), 74–83 (1993)
14. Ziegler, J.F.: Terrestrial cosmic rays and soft errors. IBM J. Res. Dev. **40**(1), 19–41 (1996)

Chapter 2
Fundamentals of Dependability

"Ah, this is obviously some strange usage of the word 'safe' that I wasn't previously aware of."
Douglas Adams, The Hitchhikers Guide to the Galaxy

The ultimate goal of fault tolerance is the development of a dependable system. In broad terms, *dependability* is the ability of a system to deliver its intended level of service to its users [16]. As computing becomes ubiquitous and penetrates our everyday lives on all scales, dependability becomes important not only for traditional safety-, mission-, and business-critical applications, but also for our society as a whole.

In this chapter, we study three fundamental characteristics of dependability: attributes, impairment, and means. Dependability *attributes* describe the properties which are required of a system. Dependability *impairments* express the reasons for a system to cease to perform its function or, in other words, the threats to dependability. Dependability *means* are the methods and techniques enabling the development of a dependable system, such as fault prevention, fault tolerance, fault removal, and fault forecasting.

2.1 Notation

Throughout the book, we use "$\cdot$" to denote the Boolean AND, "$\oplus$" to denote the Boolean XOR (exclusive-OR), and $\overline{x}$ to denote the Boolean complement of x (NOT). The arithmetic multiplication is denoted by "$\times$". With some abuse of notation, we use "+" for both, the Boolean OR and the arithmetic addition. Which meaning is intended becomes clear from the context.

E. Dubrova, *Fault-Tolerant Design*, DOI: 10.1007/978-1-4614-2113-9_2,

2.2 Dependability Attributes

The attributes of dependability express the properties which are expected from a system. Three primary attributes are reliability, availability, and safety. Other possible attributes include maintainability, testability, performability, and security [21]. Depending on the application, one or more of these attributes may be needed to appropriately evaluate a system behavior. For example, in an Automatic Teller Machine (ATM), the proportion of time in which the system is able to deliver its intended level of service (system availability) is an important measure. For a cardiac patient with a pacemaker, continuous functioning of the device is a matter of life and death. Thus, the ability of the system to deliver its service without interruption (system reliability) is crucial. In a nuclear power plant control system, the ability of the system to perform its functions correctly or to discontinue its function in a safe manner (system safety) is of uppermost importance.

2.2.1 Reliability

Reliability $R(t)$ of a system at time t is the probability that the system operates without a failure in the interval $[0, t]$, given that the system was performing correctly at time 0.

Reliability is a measure of the continuous delivery of correct service. High reliability is required in situations when a system is expected to operate without interruptions, as in the case of a heart pacemaker, or when maintenance cannot be performed because a system cannot be accessed, as in the case of deep-space applications. For example, a spacecraft mission control system is expected to provide uninterrupted service. A flaw in the system is likely to cause the destruction of the spacecraft, as happened in the case of NASA's earth-orbiting Lewis spacecraft launched on August 23, 1997 [20]. The spacecraft entered a flat spin in orbit that resulted in a loss of solar power and a fatal battery discharge. Contact with the spacecraft was lost. It was destroyed on September 28, 1997. According to the report of the Lewis Spacecraft Mission Failure Investigation, the failure occurred because the design of the attitude-control system was technically flawed and the spacecraft was inadequately monitored during its early operations phase.

Reliability is a function of time. The way in which time is specified varies substantially depending on the nature of the system under consideration. For example, if a system is expected to complete its mission in a certain period of time, like in the case of a spacecraft, time is likely to be defined as a calendar time or a number of hours. For software, the time interval is often specified in so called *natural or time units*. A natural unit is a unit related to the amount of processing performed by a software-based product, such as pages of output, transactions, jobs, or queries.

Reliability expresses the probability of success. Alternatively, we can define *unreliability* $Q(t)$ of a system at time t as the probability that the system fails in the interval $[0, t]$, given that it was performing correctly at time 0. Unreliability expresses the probability of failure. The reliability and the unreliability are related as $Q(t) = 1 - R(t)$.

2.2.2 *Availability*

Relatively few systems are designed to operate continuously without interruption and without maintenance of any kind. In many cases, we are interested not only in the probability of failure, but also in the number of failures and, in particular, in the time required to make repairs. For such applications, the attribute which we would like to maximize is the fraction of time that the system is in the operational state, expressed by availability.

Availability $A(t)$ of a system at time t is the probability that the system is functioning correctly at the instant of time t.

$A(t)$ is also referred as *point* availability, or *instantaneous* availability. Often it is necessary to determine the *interval* or *mission* availability. It is defined by

$$A(T) = \frac{1}{T}\int_0^T A(t)\mathrm{d}t. \tag{2.1}$$

$A(T)$ is the value of the point availability averaged over some interval of time T. This interval might be the life-time of a system or the time to accomplish some particular task.

Finally, it is often the case that after some initial transient effect, the point availability assumes a time-independent value. Then, the *steady-state* availability defined by

$$A(\infty) = \lim_{T\to\infty} \frac{1}{T}\int_0^T A(t)\mathrm{d}t \tag{2.2}$$

is used.

If a system cannot be repaired, the point availability $A(t)$ is equal to the system's reliability, i.e., the probability that the system has not failed between 0 and t. Thus, as T goes to infinity, the steady-state availability of a nonrepairable system goes to zero

$$A(\infty) = 0.$$

Steady-state availability is often specified in terms of *downtime per year*. Table 2.1 shows the values for the availability and the corresponding downtime.

Availability is typically used as a measure of dependability for systems where short interruptions can be tolerated. Networked systems, such as telephone switching and

Table 2.1 Availability and the corresponding downtime per year

Availability (%)	Downtime
90	36.5 days/year
99	3.65 days/year
99.9	8.76 h/year
99.99	52 min/year
99.999	5 min/year
99.9999	31 s/year

web servers, fall into this category. A telephone subscriber expects to complete a call without interruptions. However, a downtime of a few minutes a year is typically considered acceptable. Surveys show that the average expectation of an online shopper for a web page to load is 2 s [2]. This means that e-commerce web sites should be available all the time and should respond quickly even when a large number of shoppers access them simultaneously. Another example is the electrical power control system. Customers expect power to be available 24 h a day, every day, in any weather condition. A prolonged power failure may lead to health hazards and financial loss. For example, on March 11, 2001, a large industrial district, Kista, a northern suburb of Stockholm, Sweden, experienced a total power outage due to a fire in a tunnel adjacent to a power station. From 7:00 a.m. to 8:35 p.m. the following evening, 50,000 people and 700 businesses employing 30,000 people were left without power [9]. Heating, ventilation, fresh water pumps, telephones, and traffic lights stopped working. All computers, locking devices, and security systems malfunctioned. The total cost of the Kista power outage is estimated at EUR 12.8 million [15].

2.2.3 Safety

Safety can be considered as an extension of reliability, namely reliability with respect to failures that may create safety hazards. From the reliability point of view, all failures are equal. For safety considerations, failures are partitioned into *fail-safe* and *fail-unsafe* ones.

As an example, consider an alarm system. The alarm may either fail to function correctly even though a danger exists, or it may give a false alarm when no danger is present. The former is classified as a fail-unsafe failure. The latter is considered a fail-safe one. More formally, safety is defined as follows.

Safety $S(t)$ of a system at time t is the probability that the system either performs its function correctly or discontinues its operation in a fail-safe manner in the interval $[0, t]$, given that the system was operating correctly at time 0.

Safety is required in *safety-critical applications* where a failure may result in human injury, loss of life, or environmental disaster [8]. Examples are industrial controllers, trains, automobiles, avionic systems, medical systems, and military systems.

History shows that many fail-unsafe failures are caused by human mistakes. For example, the Chernobyl accident on April 26, 1986, was caused by a badly planned experiment which aimed at investigating the possibility of producing electricity from the residual energy in the turbo-generators [12]. In order to conduct the experiment, all automatic shutdown systems and the emergency core cooling system of the reactor had been manually turned off. The experiment was led by an engineer who was not familiar with the reactor facility. Due to a combination of these two factors, the experiment could not be canceled when things went wrong.

2.3 Dependability Impairments

Dependability impairments are usually defined in terms of faults, errors, or failures. A common feature of the three terms is that they give us a message that something went wrong. The difference is that, in the case of a fault, the problem occurred on the physical level; in the case of an error, the problem occurred on the computational level; in the case of a failure, the problem occurred on a system level [4].

2.3.1 Faults, Errors, and Failures

A *fault* is a physical defect, imperfection, or flaw that occurs in some hardware or software component. Examples are a short circuit between two adjacent interconnects, a broken pin, or a software bug.

An *error* is a deviation from correctness or accuracy in computation, which occurs as a result of a fault. Errors are usually associated with incorrect values in the system state. For example, a circuit or a program computed an incorrect value, or incorrect information was received while transmitting data.

A *failure* is a nonperformance of some action which is due or expected. A system is said to have a failure if the service it delivers to the user deviates from compliance with the system specification for a specified period of time [16]. A system may fail either because it does not act in accordance with the specification, or because the specification did not adequately describe its function.

Faults are reasons for errors and errors are reasons for failures. For example, consider a power plant in which a computer-controlled system is responsible for monitoring various plant temperatures, pressures, and other physical characteristics. The sensor reporting the speed at which the main turbine is spinning breaks. This fault causes the system to send more steam to the turbine than is required (error), over-speeding the turbine, and resulting in the mechanical safety system shutting down the turbine to prevent it being damaged. The system is no longer generating power (system failure, fail-safe).

The definitions of physical, computational, and system level are a bit more vague when applied to software. In the context of this book, we interpret a program code as physical level, the values of a program state as computational level, and the software system running the program as system level. For example, a bug in a program is a fault, a possible incorrect value caused by this bug is an error and a possible crash of the operating system is a system failure.

Not every fault causes an error and not every error causes a failure. This is particularly evident in the case of software. Some program bugs are hard to find because they cause failures only in very specific situations. For example, in November 1985, a $32 billion overdraft was experienced by the Bank of New York, leading to a loss of $5 million in interest. The failure was caused by an unchecked overflow of an 16-bit counter [7]. In 1994, the Intel Pentium I microprocessor was discovered to be computing incorrect answers to certain floating-point division calculations [28]. For example, dividing 5505001 by 294911 produced 18.66600093 instead of 18.66665197. The problem had occurred because of the omission of five entries in a table of 1066 values used by the division algorithm. These entries should have contained the constant 2, but because the entries were empty, the processor treated them as a zero. The manner in which a system can fail is called its *failure mode*. Many systems can fail in a variety of different modes. For example, a lock can fail to be opened or fail to be closed.

Failure modes are usually classified based on their domain (value or timing failures), on the perception of a failure by system users (consistent or inconsistent failures), and on their consequences for the environment (minor failures, major failures, and catastrophic failures) [3].

The actual or anticipated consequences of a failure are called *failure effects*. Failure effects are analyzed to identify corrective actions which need to be taken to mitigate the effect of failure on the system. Failure effects are also used in planning system maintenance and testing. In safety-critical applications, evaluation of failure effects is an essential process in the design of systems from early in the development stage to design and test [25].

2.3.2 Origins of Faults

As we mentioned in the previous section, failures are caused by errors and errors are caused by faults. Faults are, in turn, caused by numerous problems occurring at the specification, implementation, or fabrication stages of the design process. They can also be caused by external factors, such as environmental disturbances or human actions, either accidental or deliberate. Broadly, we can classify the sources of faults into four groups: incorrect specification, incorrect implementation, fabrication defects, and external factors [14].

Incorrect specification results from incorrect algorithms, architectures, or requirements. A typical example is the case where the specification requirements ignore aspects of the environment in which the system operates. The system might function correctly most of the time, but there also could be instances of incorrect performance. Faults caused by incorrect specifications are called *specification faults*. Specification faults are common, for example, in System-on-Chip (SoC) designs integrating

Intellectual Property (IP) cores because core specifications provided by the core vendors do not always contain all the details that SoC designers need. This is partly due to the intellectual property protection requirements, especially for core netlists and layouts [23].

Faults due to *incorrect implementation*, usually referred to as *design faults*, occur when the system implementation does not adequately implement the specification. In hardware, these include poor component selection, logical mistakes, poor timing, or poor synchronization. In software, examples of incorrect implementation are bugs in the program code and poor software component reuse. Software heavily relies on different assumptions about its operating environment. Faults are likely to occur if these assumptions are incorrect in the new environment. The Ariane 5 rocket accident is an example of a failure caused by a reused software component [17]. The Ariane 5 rocket exploded 37 s after lift-off on June 4, 1996, because of a software fault that resulted from converting a 64-bit floating point number to a 16-bit integer. The value of the floating point number happened to be larger than the one that can be represented by a 16-bit integer. In response to the overflow, the computer cleared its memory. The memory dump was interpreted by the rocket as an instruction to its rocket nozzles, which caused an explosion.

Many hardware faults are due to *component defects*. These include manufacturing imperfections, random device defects, and components wear-outs. Fabrication defects were the primary reason for applying fault- tolerance techniques to early computing systems, due to the low reliability of components. Following the development of semiconductor technology, hardware components became intrinsically more reliable and the percentage of faults caused by fabrication defects was greatly reduced.

The fourth cause of faults is *external factors*, which arise from outside the system boundary, the environment, the user, or the operator. External factors include phenomena that directly affect the operation of the system, such as temperature, vibration, electrostatic discharge, and nuclear or electromagnetic radiation that affect the inputs provided to the system. For instance, radiation causing a cell in a memory to flip to an opposite value is a fault caused by an external factor. Faults caused by users or operators can be accidental or malicious. For example, a user can accidentally provide incorrect commands to a system that can lead to system failure, e.g., improperly initialized variables in software. Malicious faults are the ones caused, for example, by software viruses and hacker intrusions.

Example 2.1. Suppose that you are driving somewhere in Australia's outback and your car breaks down. From your point of view, the car has failed. But what is the fault that led to the failure? Here are some of the many possibilities:

1. The designer of the car did not allow for an appropriate temperature range. This could be:

    - A specification fault if the people preparing the specification did not anticipate that temperatures of over 50 °C would be encountered.

- A design fault if the specification included a possibility of temperatures of over 50°, but the designer overlooked it.
- An implementation fault, if the manufacturer did not correctly follow the design.

2. You ignored a "Kangaroo next 100 miles" sign and crashed into a kangaroo. This would be a user fault.
3. A worker from the highway department erroneously put a "Speed Limit 100" sign instead of a "Kangaroo next 100 miles" sign. This would be an operator fault.
4. The drain valve of the radiator developed leaking problems, so the radiator coolant drained off and the engine overheated. This is a component defect due to wear-out.
5. A meteor crashed into your car. This would be an environmental fault.

2.3.3 Common-Mode Faults

A *common-mode fault* is a fault which occurs simultaneously in two or more redundant components. Common-mode faults are caused by phenomena that create dependencies between the redundant units which cause them to fail simultaneously, i.e., common communication buses or shared environmental factors. Systems are vulnerable to common-mode faults, if they rely on a single source of power, cooling, or Input/Output (I/O) bus.

Another possible source of common-mode faults is a design fault which causes redundant copies of hardware or of the same software process to fail under identical conditions. The only approach to combating common-mode design faults is design diversity. *Design diversity* is the implementation of more than one variant of the function to be performed. For computer-based applications, it is shown to be more efficient to vary a design at higher levels of abstraction [5]. For example, varying algorithms is more efficient than varying the implementation details of a design, e.g., using different program languages. Since diverse designs must implement a common system specification, the possibility for dependency always arises in the process of refining the specification. Truly diverse designs eliminate dependencies by using separate design teams, different design rules, and software tools. This issue is further discussed in Sect. 7.3.4.

2.3.4 Hardware Faults

In this section, we first consider two major classes of hardware faults: permanent and transient faults. Then, we show how different types of hardware faults can be modeled.

2.3.4.1 Permanent and Transient Faults

Hardware faults are classified with respect to fault duration into permanent, transient, and intermittent faults [3].

A *permanent fault* remains active until a corrective action is taken. These faults are usually caused by some physical defects in the hardware, such as shorts in a circuit, broken interconnections, or stuck cells in a memory.

A *transient fault* (also called *soft-error* [30]) remains active for a short period of time. A transient fault that becomes active periodically is an *intermittent fault*. Because of their short duration, transient faults are often detected through the errors that result from their propagation.

Transient faults are the dominant type of faults in today's ICs. For example, about 98 % of Random Access Memories (RAM) faults are transient faults [24]. Dynamic RAMs (DRAM) experience one single-bit transient fault per day per gigabyte of memory [26]. The causes of transient faults are mostly environmental, such as alpha particles, atmospheric neutrons, electrostatic discharge, electrical power drops, or overheating.

Intermittent faults are due to implementation flaws, ageing, and wear-out, and to unexpected operating conditions. For example, a loose solder joint in combination with vibration can cause an intermittent fault.

2.3.4.2 Fault Models

It is not possible to enumerate all the possible types of faults which can occur in a system. To make the evaluation of fault coverage possible, faults are assumed to behave according to some *fault model* [11]. A fault model attempts to describe the effect of the fault that can occur.

The most common gate-level fault model is the single stuck-at fault. A *single stuck-at fault* is a fault which results in a line in a logic circuit being permanently stuck at a logic one or zero [1]. It is assumed that the basic functionality of the circuit is not changed by the fault, i.e., a combinational circuit is not transformed to a sequential circuit, or an AND gate does not become an OR gate. Due to its simplicity and effectiveness, the single stuck-at fault is the most common fault model.

In a *multiple stuck-at fault model*, multiple lines in a logic circuit are stuck at some logic values (the same or different). A circuit with k lines can have $2k$ different single stuck-at faults and $3^k - 1$ different multiple stuck-at faults. Therefore, testing for all possible multiple stuck-at faults is obviously infeasible for large circuits.

We can test a circuit for stuck-at faults by comparing the truth table of a function $f(x_1, x_2, \dots, x_n)$ implemented by the fault-free circuit with the truth table of a function $f^\alpha(x_1, x_2, \dots, x_n)$ implemented by the circuit with a stuck-at fault α. Any assignment of input variables $(x_1, x_2, \dots, x_n)$ for which $f(x_1, x_2, \dots, x_n) \neq f^\alpha(x_1, x_2, \dots, x_n)$ is a test for the stuck-at-fault α.

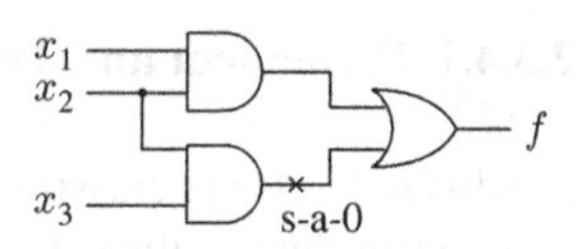

Fig. 2.1 Logic circuit for Example 2.1

Table 2.2 The truth tables of the Boolean functions implemented by the circuit in Fig. 2.1 in the fault-free case (f) and with the fault α (f^α)

x_1	x_2	x_3	f	f^α
0	0	0	0	0
0	0	1	0	0
0	1	0	0	0
0	1	1	1	0
1	0	0	0	0
1	0	1	0	0
1	1	0	1	1
1	1	1	1	1

Example 2.2. Consider the logic circuit in Fig. 2.1. It implements the Boolean function $f(x_1, x_2, x_3) = x_1x_2 + x_2x_3$. In the presence of the stuck-at-0 fault on the line marked by a cross in Fig. 2.1 (abbreviated as "s-a-0"), this function changes to $f^\alpha(x_1, x_2) = x_1x_2$. The truth tables of functions f and f^α are shown in Table 2.2. We can see that, for the input assignment $(x_1, x_2, x_3) = (0, 1, 1)$, $f(0, 1, 1) \neq f^\alpha(0, 1, 1)$. Therefore, the input assignment (0, 1, 1) is the test for this fault.

Another example of a gate-level fault model is a *bridging fault* in which two lines in a logic circuit become shorted [1]. This may result in signals being either ORed or ANDed together. A bridging fault may create a feedback loop, converting a combinational circuit to a sequential one. Bridging faults are more difficult to test because they depend upon more than one line.

2.3.5 Software Faults

Software contributes to more than a half of the system failures [10]. Specific features of software determine the main sources of its faults.

First, software does not suffer from random defects in fabrication and does not wear out. Unlike mechanical or electronic parts of hardware, software cannot be deformed, broken, or affected by environmental factors. Assuming that software is deterministic, it always behaves the same way in the same circumstances, unless there are problems in the hardware that stores the software. For these reasons, the main source of software-related failures is design faults [18]. Design faults are related to fuzzy human factors, and therefore they are harder to prevent. In hardware, design faults may also exist, but other types of faults, such as fabrication defects and transient faults caused by environmental factors, normally dominate.

Second, new faults may be introduced in software due to reliability or feature upgrades during its life cycle. A *reliability upgrade* targets enhancing reliability or security of software by redesigning or reimplementing some modules using better development processes, e.g., the cleanroom method [19]. A *feature upgrade* aims to enhance the functionality of software. However, good intentions can have bad consequences. For example, in 1991, a change of three lines of code in a several million line program caused the local telephone systems in California and along the Eastern coast to stop [13].

2.4 Dependability Means

Dependability means are the methods and techniques enabling the development of a dependable system. Fault tolerance, which is the subject of this book, is one of these methods. It is normally used in combination with other methods to attain dependability, such as fault prevention, fault removal, and fault forecasting [6]. Fault prevention aims to prevent the occurrences or introduction of faults. Fault removal aims to reduce the number of faults which are present in the system. Fault forecasting aims to estimate how many faults are present, possible future occurrences of faults, and the impact of the faults on the system.

2.4.1 Fault Tolerance

Fault tolerance targets the development of systems which function correctly in the presence of faults. As we mentioned in Sect. 1.2, fault tolerance is achieved by using some kind of redundancy. The redundancy allows a fault either to be *masked*, or *detected*, with subsequent location, containment, and recovery. Redundancy is necessary, but not sufficient for fault tolerance. For example, two duplicated components connected in parallel do not make a system fault-tolerant, unless some form of monitoring is provided, which analyzes the results and selects the correct one.

Fault masking is the process of insuring that only correct values get passed to the system output in spite of the presence of a fault. This is done by preventing the system from being affected by errors by either correcting the error, or compensating for it in some fashion. Since the system does not show the impact of the fault, the existence of the fault is invisible to the user/operator. For example, a memory protected by an error-correcting code corrects the faulty bits before the system uses the data. Another example of fault masking is triple modular redundancy with the majority voting.

Fault detection is the process of determining that a fault has occurred within a system. Examples of techniques for fault detection are acceptance tests and comparison. *Acceptance tests* is a fault detecting mechanism that can be used for systems having no replicated components. Acceptance tests are common in software systems [22]. The result of a program is subjected to a test. If the result passes the test, the program continues execution. A failed acceptance test implies a fault. *Comparison* is an alternative technique for detecting faults, used for systems with duplicated components. The output results of two components are compared. A disagreement in the results indicates a fault.

Fault location is the process of determining where a fault has occurred. A failed acceptance test cannot generally be used to locate a fault. It can only tell that something has gone wrong. Similarly, when a disagreement occurs during the comparison of two modules, it is not possible to tell which of the two has failed.

Fault containment is the process of isolating a fault and preventing the propagation of its effect throughout the system. This is typically achieved by frequent fault detection, by multiple request/confirmation protocols and by performing consistency checks between modules.

Once a faulty component has been identified, a system *recovers* by reconfiguring itself to isolate the faulty component from the rest of the system and regain operational status. This might be accomplished by having the faulty component replaced by a redundant backup component. Alternatively, the system could switch the faulty component off and continue operation with a degraded capability. This is known as *graceful degradation*.

2.4.2 Fault Prevention

Fault prevention is a set of techniques attempting to prevent the introduction or occurrence of faults in the system in the first place. Fault prevention is achieved by quality control techniques during the specification, implementation, and fabrication stages of the design process. For hardware, this includes design reviews, component screening, and testing [27]. For software, this includes structural programming, modularization, and formal verification techniques [29].

A rigorous design review may eliminate many specification faults. If a design is efficiently tested, many of its faults and component defects can be avoided. Faults introduced by external disturbances such as lightning or radiation are prevented by shielding, radiation hardening, etc. User and operation faults are avoided by training and regular procedures for maintenance. Deliberate malicious faults caused by viruses or hackers are reduced by firewalls or similar security means.

2.4.3 Fault Removal

Fault removal is a set of techniques targeting the reduction of the number of faults which are present in the system. Fault removal is performed during the development phase as well as during the operational life of a system. During the development phase, fault removal involves three steps: verification, diagnosis, and correction.

Fault removal during the operational life of the system consists of corrective and preventive maintenance.

Verification is the process of checking whether the system meets a set of given conditions. If it does not, the other two steps follow: the fault that prevents the conditions from being fulfilled is diagnosed and the necessary corrections are performed.

In *preventive maintenance*, parts are replaced, or adjustments are made before failure occurs. The objective is to increase the dependability of the system over the long term by staving off the ageing effects of wear-out. In contrast, *corrective maintenance* is performed after the failure has occurred in order to return the system to service as soon as possible.

2.4.4 Fault Forecasting

Fault forecasting is a set of techniques aiming to estimate how many faults are present in the system, possible future occurrences of faults, and the consequences of faults. Fault forecasting is done by performing an evaluation of the system behavior with respect to fault occurrences or activation. The evaluation can be *qualitative*, which aims to rank the failure modes or event combinations that lead to system failure, or *quantitative*, which aims to evaluate in terms of probabilities the extent to which some attributes of dependability are satisfied. Simplistic estimates merely measure redundancy by accounting for the number of redundant success paths in a system. More sophisticated estimates account for the fact that each fault potentially alters a system's ability to resist further faults. We study qualitative evaluation techniques in more detail in the next chapter.

2.5 Summary

In this chapter, we have studied the basic concepts of dependability and their relationship to fault tolerance. We have considered three primary attributes of dependability: reliability, availability, and safety. We have analyzed various reasons for a system to cease to perform its function. We have discussed complementary techniques to fault tolerance enabling the development of a dependable system, such as fault prevention, fault removal, and fault forecasting.

Problems

2.1 What is the primary goal of fault tolerance?

2.2 Give three examples of applications in which a system failure can cost people's lives or environmental disaster.

2.3 What is the dependability of a system? Why is the dependability of computing systems a critical issue nowadays?
2.4 Describe three fundamental characteristics of dependability.
2.5 What do the attributes of dependability express? Why are different attributes used in different applications?
2.6 Define the reliability of a system. What property of a system does the reliability characterize? In which situations is high reliability required?
2.7 Define point, interval and steady-state availabilities of a system. Which attribute would we like to maximize in applications requiring high availability?
2.8 What is the difference between the reliability and the availability of a system? How does the point availability compare to the system's reliability if the system cannot be repaired? What is the steady-state availability of a non-repairable system?
2.9 Compute the downtime per year for $A(\infty) = 90, 75$, and $50\,\%$.
2.10 A telephone system has less than 3 min per year downtime. What is its steady-state availability?
2.11 Define the safety of a system. Into which two groups are the failures partitioned for safety analysis? Give examples of applications requiring high safety.
2.12 What are dependability impairments?
2.13 Explain the differences between faults, errors, and failures and the relationships between them.
2.14 Describe the four major groups of fault sources. Give an example for each group. In your opinion, which of the groups causes the "most expensive" faults?
2.15 What is a common-mode fault? By what kind of phenomena are common-mode faults caused? Which systems are most vulnerable to common-mode faults? Give examples.
2.16 How are hardware faults classified with respect to fault duration? Give an example for each type of fault.
2.17 Why are fault models introduced? Can fault models guarantee $100\,\%$ accuracy?
2.18 Give an example of a combinational logic circuit in which a single stuck-at fault on a given line never causes an error on the output.
2.19 Consider the logic circuit of a full adder shown on Fig. 5.11.

1. Find a test for stuck-at-1 fault on input b.
2. Find a test for stuck-at-0 fault on the fan-out branch of input a which feeds into an AND gate.

2.20 Prove that a logic circuit with k lines can have $3^k - 1$ different multiple stuck-at faults.
2.21 Suppose that we modify the stuck-at fault model in the following way. Instead of having a line in a logic circuit being permanently stuck at a logic one or zero value, we have a transistor being permanently open or closed. Draw a transistor-level circuit diagram of a CMOS NAND gate.

1. Give an example of a fault in your circuit which can be modeled by the new model but cannot be modeled by the standard stuck-at fault model.

2. Find a fault in your circuit which cannot be modeled by the new model but can be modeled by the standard stuck-at fault model.

2.22 Explain the main differences between software and hardware faults.
2.23 What are dependability means? What are the primary goals of fault prevention, fault removal, and fault forecasting?
2.24 What is redundancy? Is redundancy necessary for fault tolerance? Is any redundant system fault-tolerant?
2.25 Does a fault need to be detected to be masked?
2.26 Define fault containment. Explain why fault containment is important.
2.27 Define graceful degradation. Give an example of application where graceful degradation is desirable.
2.28 How is fault prevention achieved? Give examples for hardware and for software.
2.29 During which phases of a system's life is fault removal performed?
2.30 What types of faults are targeted by verification?
2.31 What are the objectives of preventive and corrective maintenance?

References

1. Abramovici, M., Breuer, M.A., Frideman, A.D.: Digital system testing and testable design. Computer Science Press, New York (1995)
2. Akamai: Akamai reveals 2 seconds as the new threshold of acceptability for ecommerce web page response times (2000). http://www.akamai.com/html/about/press/releases/2009/press_091409.html
3. Avižienis, A.: Fault-tolerant systems. IEEE Trans. Comput. **25**(12), 1304–1312 (1976)
4. Avižienis, A.: The four-universe information system model for the study of fault-tolerance. In: Proceedings of the 12th Annual International Symposium on Fault-Tolerant Computing, FTCS'82, IEEE Press, pp. 6–13 (1982)
5. Avižienis, A.: Design diversity: An approach to fault tolerance of design faults. In: Proceedings of the National Computer Conference and Exposition, pp. 163–171 (1984)
6. Avizienis, A., Laprie, J.C., Randell, B., Landwehr, C.: Basic concepts and taxonomy of dependable and secure computing. IEEE Trans. Dependable Secur. Comput. **1**(1), 11–33 (2004)
7. Berry, J.M.: $32 billion overdraft resulted from snafu (1985). http://catless.ncl.ac.uk/Risks/1.31.html#subj4
8. Bowen, J., Stravridou, V.: Safety-critical systems, formal methods and standards. IEE/BCS Softw. Eng. J. **8**(4), 189–209 (1993)
9. Deverell, E.: The 2001 Kista Blackout: Corporate Crisis and Urban Contingency. The Swedish National Defence College, Stockholm (2003)
10. Gray, J.: A census of TANDEM system availability between 1985 and 1990. IEEE Trans. Reliab. **39**(4), 409–418 (1990)
11. Hayes, J.: Fault modeling for digital MOS integrated circuits. IEEE Trans. Comput. Aided Des. Integr. Circuits Syst. **3**(3), 200–208 (1984)
12. IAEA: Frequently asked Chernobyl questions (2005). http://www.iaea.org/newscenter/features/chernobyl-15/cherno-faq.shtml
13. Joch, A.: How software doesn't work: nine ways to make your code reliable (1995). http://www.welchco.com/02/14/01/60/95/12/0102.HTM
14. Johnson, B.W.: The Design and Analysis of Fault Tolerant Digital Systems. Addison-Wesley, New York (1989)

15. Karlsson, I.: Utvärdering av birka energi (Birka Energi's Evaluation), Sweden (2001)
16. Laprie, J.C.: Dependable computing and fault tolerance: Concepts and terminology. In: Proceedings of 15th International Symposium on Fault-Tolerant Computing (FTSC-15), IEEE Computer Society, pp. 2–11 (1985)
17. Lions, J.L.: Ariane 5 flight 501 failure, report by the inquiry board (1996). http://www.esrin.esa.it/htdocs/tidc/Press/Press96/ariane5rep.html
18. Lyu, M.R.: Introduction. In: Lyu, M.R. (ed.) Handbook of Software Reliability, pp. 3–25. McGraw-Hill, New York (1996)
19. Mills, H., Dyer, M., Linger, R.: Cleanroom software engineering. IEEE Softw. **4**(5), 19–25 (1987)
20. NASA: The Role of Small Satellites in NASA and NOAA Earth Observation Programs. Space Studies Board, National Research Council, National Academy of Sciences, Washington, USA (2000)
21. Nelson, V.P.: Fault-tolerant computing: fundamental concepts. IEEE Comput. **23**(7), 19–25 (1990)
22. Randell, B.: System structure for software fault tolerance. In: Proceedings of the International Conference on Reliable Software, pp. 437–449 (1975)
23. Saleh, R., Wilton, S., Mirabbasi, S., Hu, A., Greenstreet, M., Lemieux, G., Pande, P., Grecu, C., Ivanov, A.: System-on-chip: Reuse and integration. Proc. IEEE **94**(6) (2006)
24. Smith, M.: RAM reliability: Soft errors (1998). http://www.crystallineconcepts.com/ram/ram-soft.html
25. Smith, M.D.J., Simpson, K.G.: Safety Critical Systems Handbook, 3rd edn. Elsevier Ltd., New York (2011)
26. Tezzaron Semiconductor: Soft errors in electronic memory (2004). http://www.tezzaron.com/about/papers/papers.html
27. Tumer, I.Y.: Design methods and practises for fault prevention and management in spacecraft. Tech. Rep. 20060022566, NASA (2005)
28. Pratt, V.: Anatomy of the pentium bug. In: Mosses, P.D., Nielsen, M., Schwartzbach, M.I. (eds.) TAPSOFT'95: Theory and Practice of Software Development, vol. 915, pp. 97–107. Springer, Verlag (1995)
29. Yu, W.D.: A software fault prevention approach in coding and root cause analysis. Bell Labs Tech. J. **3**(2), 3–21 (1998)
30. Ziegler, J.F.: Terrestrial cosmic rays and soft errors. IBM J. Res. Dev. **40**(1), 19–41 (1996)

Chapter 3
Dependability Evaluation Techniques

"A common mistake that people make when trying to design something completely foolproof is to underestimate the ingenuity of complete fools."

Douglas Adams, Mostly Harmless

Along with cost and performance, dependability is the third critical criterion upon which system-related decisions are made. Dependability evaluation is important, because it helps identify aspects of the system which are critical for its dependability. Such aspects can be, for example, component reliability, fault coverage, or maintenance strategy. Once the critical points are identified, design engineers can focus on their improvements early in the product development stage.

There are two conventional approaches to dependability evaluation: (1) modeling of a system at the design phase, and (2) assessment of the system in a later phase, typically by test. The first approach relies on models that use component-level failure rates available from handbooks or supplied by the manufacturers to evaluate the overall system dependability. Such an approach provides an early estimate of system dependability, which is its great advantage. However, both the model and the underlying assumptions need to be validated by actual measurements later in the design process.

The second approach typically uses test data and reliability growth models [1, 3]. It involves fewer assumptions than the first, but it can be very costly. The higher the dependability required for a system, the longer the test. A further difficulty arises in the translation of reliability data obtained by test into those applicable in the operational environment. In this chapter, we focus on the first approach to dependability evaluation.

Dependability evaluation can be qualitative or quantitative. The *qualitative evaluation* aims to identify, classify, and rank the failure modes, or the event combinations that can lead to system failures. For example, component faults or environmental conditions can be analyzed. The *quantitative evaluation* aims to evaluate reliability, availability, or the safety of a system in terms of probabilities. This can be done

E. Dubrova, *Fault-Tolerant Design*, DOI: 10.1007/978-1-4614-2113-9_3,

using, for example, reliability block diagrams or Markov processes. In this chapter, we focus on the quantitative evaluation.

The chapter is organized as follows. We begin with a brief introduction to probability theory, necessary for understanding the presented material. Then, we introduce common measures of dependability, such as failure rate, mean time to failure, mean time to repair, etc. Afterwards, we describe combinatorial and stochastic dependability models. Finally, we show how these models can be used for evaluating system reliability, availability, and safety.

3.1 Basics of Probability Theory

Probability is the branch of mathematics which studies the possible outcomes of given events together with their relative likelihoods and distributions [5]. In common language, the word "probability" is used to mean the chance of a particular event occurring expressed on a scale from 0 (impossibility) to 1 (certainty).

The first axiom of probability theory states that the value of probability of an event A belongs to the interval [0,1]:

$$0 \leq P(A) \leq 1. \tag{3.1}$$

Let $\overline{A}$ denote the event "not A". For example, if A stands for "it rains", $\overline{A}$ stands for "it does not rain". The second axiom of probability theory says that the probability of an event $\overline{A}$ is equal to 1 minus the probability of the event A:

$$P(\overline{A}) = 1 - P(A). \tag{3.2}$$

Suppose that one event, A is dependent on another event, B. Then $P(A|B)$ denotes the conditional probability of event A, given event B. The fourth rule of probability theory states that the probability $P(A \cdot B)$ that both A and B occur is equal to the probability that B occurs times the conditional probability $P(A|B)$:

$$P(A \cdot B) = P(A|B) \times P(B), \text{ if } A \text{ depends on } B. \tag{3.3}$$

The notation $P(A|B)$ reads as "probability of A given B".

If $P(B) > 0$, then the Eq. (3.3) can be re-written as

$$P(A|B) = \frac{P(A \cdot B)}{P(B)}. \tag{3.4}$$

A special case is the situation in which two events are mutually independent. If events A and B are independent, then the probability $P(A)$ does not depend on whether B has already occurred or not, and vice versa. Hence, $P(A|B) = P(A)$.

Thus, for independent events, the Eq. (3.3) reduces to

$$P(A \cdot B) = P(A) \times P(B), \text{ if } A \text{ and } B \text{ are independent events.} \quad (3.5)$$

Another special case is the situation in which two events are mutually exclusive. This means that, if A occurs, B cannot, and vice versa. In this case, $P(A \cdot B) = 0$ and $P(B \cdot A) = 0$. Therefore, the Eq. (3.3) becomes

$$P(A \cdot B) = 0, \text{ if } A \text{ and } B \text{ are mutually exclusive events.} \quad (3.6)$$

Finally, consider the case when either A, or B, or both events may occur. Then, the probability $P(A + B)$ is given by

$$P(A + B) = P(A) + P(B) - P(A \cdot B). \quad (3.7)$$

Combining (3.6) and (3.7), we get

$$P(A + B) = P(A) + P(B), \text{ if } A \text{ and } B \text{ are mutually exclusive events.} \quad (3.8)$$

Given n independent events which can either be a success or a failure, the probability that exactly k out of n of events are a success is given by

$$P(k \text{ out of } n \text{ success}) = \binom{n}{k} P^k (1 - P)^{n-k},$$

where P is the probability that a single event is a success and

$$\binom{n}{k} = \frac{n!}{(n-k)!k!}.$$

The probability that at least k out of n events are a success is given by

$$P(\geq k \text{ out of } n \text{ success}) = \sum_{i=k}^{n} \binom{n}{i} P^i (1 - P)^{n-i}. \quad (3.9)$$

Example 3.1. Compute the probability that no more than two out of four tires in a four-wheel car fail, assuming that the reliability of a single tire is $R(t)$ and that the failures of the individual tires are independent events.

By rule (3.2), the probability that no more than two out of four tires fail is given by

$$P(\leq 2 \text{ out of 4 fail}) = 1 - P(\geq 3 \text{ out of 4 work}).$$

By Eq. (3.9), the probability that three or more tires work is given by

$$\begin{aligned} P(\geq 3\,\text{out of}\,4\,\text{work}) &= \binom{4}{3} R(t)^3(1-R(t)) + \binom{4}{4} R(t)^4 \\ &= 4R(t)^3(1-R(t)) + R(t)^4. \end{aligned}$$

3.2 Common Measures of Dependability

In this section, we describe common dependability measures such as failure rate, mean time to failure, mean time to repair, mean time between failures, and fault coverage.

3.2.1 Failure Rate

Failure rate is defined as the expected number of failures per unit time [15]. For hardware, the typical evolution of failure rate over the lifetime of a system is illustrated by the *bathtub curve* shown in Fig. 3.1. This curve has three phases [15]: (I) infant mortality, (II) useful life, and (III) wear out. In the beginning of the system's life, the failure rate sharply decreases due to frequent failures of weak components whose manufacturing defects were overlooked during manufacturing testing. After a certain time, the failure rate tends to stabilize and remains constant. Finally, as electronic or mechanical components wear out, the failure rate increases.

During the useful life phase of the system, the failure rate function $z(t)$ is assumed to have a constant value λ. Then, the reliability of the system decreases exponentially with time:

$$R(t) = e^{-\lambda t}. \tag{3.10}$$

This law is known as the *exponential failure law*. The plot illustrating it is shown in Fig. 3.2.

Failure rate data are often available at component level. There are government and commercial organizations which collect and publish failure rate estimates for

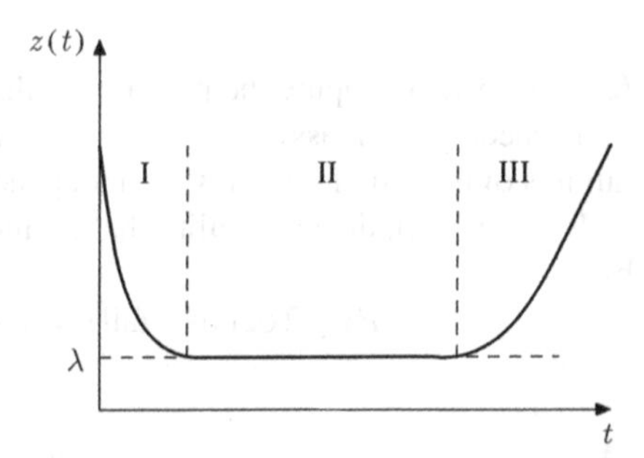

Fig. 3.1 Typical evolution of failure rate in a hardware system: *I* infant mortality phase, *II* useful life, *III* wear-out phase

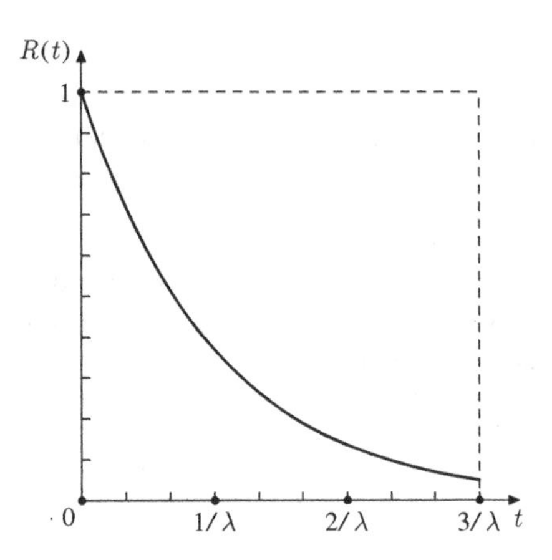

Fig. 3.2 Plot illustrating the exponential failure law

various components and parts. For example, IEEE Standard 500 provides failure rates for electrical, electronic and sensing components, as well as mechanical equipment for nuclear-power generating stations [6]; the military standard MIL-HDBK-338 [8] describes failure rates for military electronic components. Nonelectronic component failure rates are available from [10]. The offshore reliability data (OREDA) database presents statistical analysis on many types of process equipment [11].

If failure rates of components are available, a crude estimate of the failure rate of a system without redundancy during its useful life phase can be obtained by summing up the failure rates of the components:

$$\lambda = \sum_{i=1}^{n} \lambda_i \tag{3.11}$$

where λ_i is the failure rate of the component i, for $i \in \{1, 2, \ldots, n\}$. The above equation assumes that component failures are independent events.

Example 3.2. A logic circuit with no redundancy consists of 16 two-input NAND gates and three flip-flops. NAND gates and flip-flops have constant failure rates of 0.311×10^{-7} and 0.412×10^{-7} per hour, respectively. Assuming that component failures are independent events, compute:

1. The failure rate of the logic circuit.
2. The reliability of the logic circuit for a 1-year mission.

(1) Since the logic circuit contains no redundant components, from the Eq. (3.11), we can conclude that the failure rate of the logic circuit is:

$$\lambda_{\text{circuit}} = 16\lambda_{\text{AND}} + 3\lambda_{\text{FF}} = 0.6212 \times 10^{-6} \text{ per hour.}$$

(2) The reliability of the logic circuit for a 1-year mission can be computed from Eq. (3.10) as

$$R(t) = e^{-\lambda t} = e^{-0.6212\times 10^{-6}\times 8760} = 0.9946.$$

So, the reliability of the logic circuit for a 1-year mission is 99.46 %. Note that we converted 1 year into hours as 1 year = 365 days × 24 hours = 8760 hours.

Example 3.3. A car manufacturer estimates that the reliability of their gear boxes is 82 % during the first 3 years of operation. Assuming constant failure rate, determine how many cars will be returned for repair during the first year of operation due to gear box failure.

From Eq. (3.10), we can derive the failure rate of a gear box as follows:

$$\lambda = -\frac{\ln R(t)}{t} = -\frac{\ln 0.82}{3} = 0.0662 \text{ per year.}$$

We can conclude that 6.62 % of cars will be returned for repair during the first year of operation due to a gear-box failure.

Example 3.4. An ATM manufacturer determines that their product has a constant failure rate of $\lambda = 0.017$ per year in normal use. Estimate for how long a warranty should be set if no more than 5 % of machines are to be returned for repair.

If no more than 5 % of machines are to be returned for repair, the reliability should be $R(t) \geq 0.95$. From Eq. (3.10), we can derive the interval of time during which $R(t) \geq 0.95$ as follows:

$$t \leq -\frac{\ln R(t)}{\lambda} = -\frac{\ln 0.95}{0.017} = 3.0172 \text{ years.}$$

We can conclude that the warranty should be set to 3 years.

The exponential failure law is useful for the reliability evaluation of hardware systems, for which the assumption of constant failure rate is realistic. However, in software systems, failure rate usually decreases as a function of time. Time-varying failure rate functions can be modeled using *Weibull distribution* [15]:

$$z(t) = \alpha\lambda(\lambda t)^{\alpha-1}$$

where α and λ are constants determining the behavior of $z(t)$ over time. We can distinguish the following three cases:

1. If $\alpha = 1$, then $z(t) = \lambda$.
2. If $\alpha > 1$, then $z(t)$ increases with time.
3. If $\alpha < 1$, then $z(t)$ decreases with time.

So, software systems can be modeled using the Weibull distribution with $\alpha < 1$ [7].

A typical evolution of failure rate in a software system is shown in Fig. 3.3. The three phases of the evolution are [13]: (I) test/debug, (II) useful life (II), and (III) obsolescence.

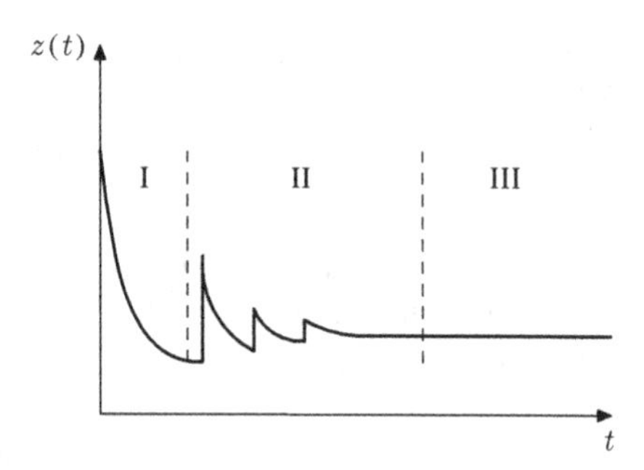

Fig. 3.3 Typical evolution of failure rate in a software system: *I* test/debug phase, *II*, useful life, *III* obsolescence

There are two major differences between hardware and software failure rate evolution [13]. One is that, in the useful-life phase, software usually experiences periodic increases in failure rate due to feature upgrades. Since a feature upgrade enhances the functionality of software, it also increases its complexity, which makes the probability of faults higher. After a feature upgrade, the failure rate levels off gradually, due to the bugs found and fixed. The second difference is that, in the last phase, the software failure rate does not increase. In this phase, the software is approaching obsolescence. Therefore, there is no motivation for more upgrades or changes.

Software failure rate during useful life depends on many factors, including complexity and size of code, software process used to develop the code, experience of the development team, percentage of code reused from a previous stable project, and rigor and depth of testing at test/debug (I) phase.

3.2.2 Mean Time to Failure

Another frequently used dependability measure is mean time to failure, defined as follows.

Mean Time to Failure (MTTF) of a system is the expected time of the occurrence of the first system failure.

If we put n identical components into operation at the time $t = 0$ and we measure the time t_i that each component i operates before failing, for $i = \{1, 2, \ldots, n\}$, then MTTF is given by:

$$\text{MTTF} = \frac{1}{n}\sum_{i=1}^{n} t_i. \tag{3.12}$$

MTTF is related to the reliability as follows:

$$\text{MTTF} = \int_{0}^{\infty} R(t)\text{d}t. \tag{3.13}$$

If the reliability function obeys the exponential failure law, then the solution of (3.13) is given by

$$\text{MTTF} = \frac{1}{\lambda}. \tag{3.14}$$

It is common to present MTTF in *Failures In Time* (FIT) format, which shows how many failures can be expected from one billion hours of operation. If MTTF is expressed in hours, then

$$\text{FIT} = \frac{10^9}{\text{MTTF}}.$$

Example 3.5. A manufacturer of electrical motors determines that their product has a constant failure rate of $\lambda = 0.549 \times 10^{-3}$ per hour in normal use. Compute the MTTF and FIT of the product.

From Eq. (3.14), we can compute MTTF as follows:

$$\text{MTTF} = \frac{1}{\lambda} = \frac{1}{0.549 \times 10^{-3}} = 1822 \text{ h}.$$

By re-formatting MTTF per billion hours of operation, we get the following FIT rate:

$$\text{FIT} = \frac{10^9}{\text{MTTF}} = \frac{10^9}{1822} = 548847.$$

3.2.3 Mean Time to Repair

Mean Time to Repair (MTTR) of a system is the average time required to repair the system.

MTTR is commonly specified in terms of the *repair rate* μ, which is the expected number of repairs per unit of time [16]:

$$\text{MTTR} = \frac{1}{\mu}. \tag{3.15}$$

MTTR depends on the fault recovery mechanism used in the system, the location of the system, the location of spare modules (on-site versus off-site), the maintenance schedule, and so on. A low MTTR requirement means the system has a high operational cost. For example, if repair is done by replacing the hardware module, the hardware spares are kept on-site and the site is maintained 24 h a day, then the expected MTTR can be less than an hour. However, if the site maintenance is relaxed to regular working hours on week days only, the expected MTTR can increase to several days. If the system is remotely located and the operator needs to be flown in to replace the faulty module, the MTTR can be several weeks. In software, if the failure is detected by watchdog timers and the processor automatically restarts the failed tasks, without operating system reboot, then MTTR can be less than a minute. If software fault detection is not supported and a manual reboot by an operator is required, then the MTTR can range from minutes to weeks, depending on the location of the system.

Example 3.6. Compute what repair rates per hour should be set to achieve the following MTTR: (a) 30 min, (b) 3 days, and (c) 2 weeks.

From the Eq. (3.15), we can conclude that the repair rates should be set to: (a) $\mu = 2$ per hour, (b) $\mu = 0.1389 \times 10^{-1}$ per hour, (c) $\mu = 0.2976 \times 10^{-2}$ per hour.

If the system experiences n failures during its lifetime, then the total time that the system is operational is $n \times$ MTTF. Similarly, the total time that the system is repaired is $n \times$ MTTR. Thus, the expression (2.2) for steady-state availability can be approximated as [16]:

$$A(\infty) = \frac{n\,\text{MTTF}}{n\,\text{MTTF} + n\,\text{MTTR}} = \frac{\text{MTTF}}{\text{MTTF} + \text{MTTR}}. \tag{3.16}$$

By expressing MTTF and MTTR according to the Eqs. 3.15 and 3.14, we get

$$A(\infty) = \frac{\mu}{\mu + \lambda}. \tag{3.17}$$

Example 3.7. A printer has an MTTF = 2160 h and MTTR = 36 h.

1. Estimate its steady-state availability.
2. Compute what MTTF can be tolerated without decreasing the steady-state availability of the printer if MTTR is reduced to 12 h.

(1) From Eq. 3.16, we can conclude that the steady-state availability of the printer is

$$A(\infty) = \frac{\text{MTTF}}{\text{MTTF} + \text{MTTR}} = \frac{2160}{2196} = 0.9836.$$

(2) From Eq. 3.16, we can derive the following formula of MTTF as a function of $A(\infty)$ and MTTR:

$$\text{MTTF} = \frac{A(\infty) \times \text{MTTF}}{1 - A(\infty)}.$$

If MTTR = 12 and $A(\infty) = 0.9836$, we get

$$\text{MTTF} = \frac{0.9836 \times 12}{1 - 0.9836} = 719.7\text{ h}.$$

So, if MTTR is reduced to 12 h, then we can tolerate MTTF = 719.7 h without decreasing the steady-state availability of the printer.

Example 3.8. A copy machine has a constant failure rate of 0.476×10^{-3} per hour. Estimate what repair rate should be maintained to achieve a steady-state availability of 98.87 %.

From Eq. 3.17, we can derive the following formula of repair rate as a function of $A(\infty)$ and failure rate:

$$\mu = \frac{\lambda}{(1 - A(\infty))} = \frac{0.476 \times 10^{-3}}{1 - 0.9887} = 0.0421 \text{ per hour}.$$

3.2.4 Mean Time Between Failures

Mean Time Between Failures (MTBF) of a system is the average time between failures of the system.

If we assume that a repair makes a system perfect, then MTBF and MTTF are related as follows:

$$\text{MTBF} = \text{MTTF} + \text{MTTR}. \tag{3.18}$$

Example 3.9. A telephone switching system has a constant failure rate of 0.12×10^{-2} per hour and a downtime of 3 min per year. Estimate its MTBF.

If a system has a downtime of 3 min per year, then its steady-state availability is $A(\infty) = 0.9994$. The MTTF of the system is $\text{MTTF} = 1/\lambda = 833.3$ h. From the Eqs. 3.16 and 3.18, we can conclude that

$$\text{MTBF} = \frac{\text{MTTF}}{A(\infty)} = \frac{833.3}{0.9994} = 833.8.$$

3.2.5 Fault Coverage

There are several types of fault coverage, depending on whether we are concerned with fault detection, fault location, fault containment, or fault recovery. Intuitively, fault coverage is the probability that the system does not fail to perform an expected actions when a fault occurs. More formally, different types of fault coverage are defined in terms of the conditional probability $P(A|B)$ as follows [7].

Fault-detection coverage is the conditional probability that, given the existence of a fault, the system detects it:

$$C = P(\text{fault detection}|\text{fault existence}).$$

For example, a typical industrial requirement is that 99 % of detectable single stuck-at faults are detected during manufacturing test of application specific integrated circuits (ASICs) [12]. The fault-detection coverage $C = 0.99$ can be used as a measure of the system's ability to meet such a requirement.

In a similar way, we can define other types of fault coverage. *Fault location coverage* is the conditional probability that, given the existence of a fault, the system locates it:

$$C = P(\text{fault location}|\text{fault existence}).$$

Fault containment coverage is the conditional probability that, given the existence of a fault, the system contains it.

$$C = P(\text{fault containment}|\text{fault existence}).$$

Fault recovery coverage is the conditional probability that, given the existence of a fault, the system recovers.

$$C = P(\text{fault recovery}|\text{fault existence}).$$

Example 3.10. A logic circuit with 3,200 lines has 20 undetectable stuck-at faults. The test set developed for its manufacturing testing is capable of detecting 6,252 single stuck-at faults in the circuit. Check if the resulting fault coverage meets the industrial requirement of 99 % fault-detection coverage for detectable faults.

Since there are two possible faults per each line, stuck-at-0 and stuck-at-1, the total number of faults in the circuit is 6,400. The fault-detection coverage for detectable faults can be computed as:

$$C = \frac{\text{Number of detected faults}}{\text{Number of detectable faults}} = \frac{6252}{6400 - 20} = 0.9799.$$

We can conclude that the developed test set does not meet the industrial requirement of 99 % fault-detection coverage for detectable faults.

3.3 Dependability Modeling

In this section we consider two classes of dependability models: combinatorial and stochastic. *Combinatorial models* assume that the failures of individual components are mutually independent. *Stochastic processes* take the dependencies between component's failures into account, enabling analysis of more complex scenarios.

3.3.1 Reliability Block Diagrams

Combinatorial reliability models include reliability block diagrams, fault trees, and reliability graphs. In this book, we focus on reliability block diagrams, which are

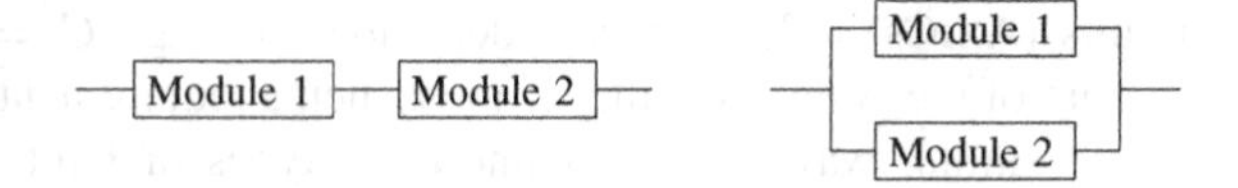

Fig. 3.4 RBDs of two-component serial (*left*) and parallel (*right*) systems

the oldest and most common reliability model. We mention fault trees and reliability graphs only briefly.

A reliability block diagram (RBD) presents an abstract view of a system. The components are represented as blocks. The interconnections among the blocks show the operational dependency among the components of a system. They are not necessarily related to the physical connection of components. Blocks are connected in *series* if all of them are necessary for the system to be operational. Blocks are connected in *parallel* if only one of them is sufficient for the system to be operational. RBDs of two-component serial and parallel systems are shown in Fig. 3.4. Models of more complex systems can be built by partitioning a system into serial and parallel subsystems and combining the RBDs of these subsystems into one.

As an example, consider a system consisting of two duplicated processors and a memory. The RBD of this system is shown in Fig. 3.5. The processors are connected in parallel, since only one of them is sufficient for the system to be operational. The memory is connected in series, since its failure would cause system failure.

Example 3.11. A system with five modules: A, B, C, D, and E is connected, so that it operates correctly if (1) either modules A and D operate correctly, or (2) module C operates correctly and either D or E operates correctly. Draw an RBD of this system.

An RBD of the system is shown in Fig. 3.6.

RBDs are a popular model, because they are easy to understand and to use for modeling systems with redundancy. In the next section, we show that they are also easy to evaluate using analytical methods. However, RBDs, as well as other combinatorial reliability models, have a number of limitations. First, RBDs assume that the states of components are limited to operational and failed and that a system's configuration does not change during its lifetime. For this reason, they cannot model standby components as well as complex fault detection and repair mechanisms. Second, the failures of individual components are assumed to be independent events. Therefore, the effects due to load sharing cannot be adequately represented by RBDs.

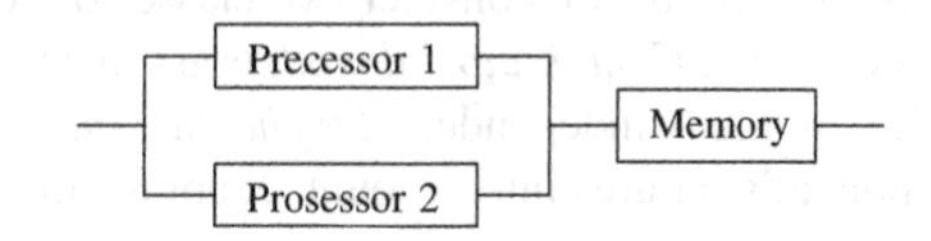

Fig. 3.5 RBD of a three-component system consisting of two duplicated processors and a memory

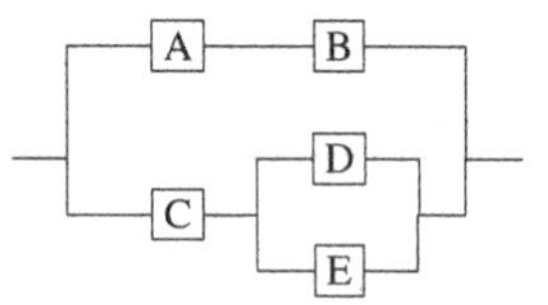

Fig. 3.6 An RBD of the system from Example 3.11

3.3.2 Fault Trees

The fault-tree model was developed by Bell Telephone Laboratories to perform safety evaluation of the Minuteman Launch Control System [4]. Later, it was applied to reliability analysis.

Fault trees describe possible sequences of events that can lead to a system failure. The operational dependency between the components of a system is represented using Boolean gates (e.g. AND, OR). Components are represented as input nodes in the tree, providing inputs to the gates. When a component fails, the corresponding input to the gate becomes TRUE. If any input of an OR gate becomes TRUE, then its output also becomes TRUE. The inputs of an OR gate are those components which are required to be operational for the system to be operational. The inputs of an AND gate, on the other hand, are those components all of which should fail for the system to fail. The output of an AND gate becomes TRUE only if all its inputs become TRUE. When the output of the gate at the root of the tree becomes TRUE, the system is considered failed. System reliability is calculated by converting a fault tree into an equivalent set of Boolean equations [17].

An RBD may be converted into a fault tree by replacing series paths with AND gates and parallel paths with OR gates and then applying de Morgan's rules to the resulting tree.

3.3.3 Reliability Graphs

A *reliability graph* [2] is an acyclic directed graph. Each component of a system is represented by a directed edge between two vertices. The failure in the component is modeled by deleting the edge from the graph. Some special edges, called ∞-edges, represent components that do not fail. The graph has two designated vertices, called *source* and *sink*. The source vertex has no incoming edges. The sink vertex has no outgoing edges. The system is considered operational as long as there exists at least one directed path from the source vertex to the sink vertex.

Reliability graphs are a simple and intuitive model, because their structure matches the structure of the modeled system. However, due to their limited expression power, for complex systems drawing reliability graphs becomes too difficult to be practical.

Table 3.1 Four types of Markov processes

State space	Time space	Common model name
Discrete	Discrete	Discrete time Markov chains
Discrete	Continuous	Continuous time Markov chains
Continuous	Discrete	Continuous state, discrete time Markov processes
Continuous	Continuous	Continuous state, continuous time Markov processes

3.3.4 Markov Processes

The theory of Markov processes derives its name from the Russian mathematician A. A. Markov (1856–1922), who was the first to describe stochastic processes formally.

Markov processes are a special class of stochastic processes [9]. The basic assumption is that the behavior of the system in each state is *memoryless*. This means that the transition from the current state of a system is determined only by this state and not by the previous states or the time at which the present state is reached.

Before a transition occurs, the time spent in each state follows an exponential distribution. This assumption is satisfied if all events (failures, repairs, etc.) occur at a constant occurrence rate. Due to this assumption, Markov processes cannot be used to model the behavior of systems that are subjected to component wear outs. General stochastic processes should be used instead.

Markov processes are classified according to state space and time-space characteristics as shown in Table 3.1.

In most dependability analysis applications, the state space is discrete. For example, each component of a system may have two states: operational or failed. The time scale is usually continuous, which means that component failure and repair times are random variables. Thus, *Continuous Time Markov Chains* (also called *Continuous Markov Models*) are most common. In this book, we focus on this type of Markov chains. There are, however, applications in which the time scale is discrete. Examples are synchronous communication protocol and shifts in equipment operation. If both time and state space are discrete, then the process is called a *discrete time Markov chain*.

Markov processes are illustrated graphically by state transition diagrams. A *state transition diagram* is a directed graph $G = (V, E)$, where V is the set of vertices representing *system states* and E is the set of edges representing *system transitions*. For dependability models, a state is defined to be a particular combination of operating and failed components. For example, if we have a system consisting of two components, then there are four different combinations enumerated in Table 3.2, where O indicates an operational component and F indicates a failed component.

The state transitions reflect changes which occur within the system state. For example, if a system with two identical components is in the state $(O\,O)$ and the first

Table 3.2 States of a Markov chain of a two-component system

Component		State
1	2	Number
O	O	1
O	F	2
F	O	3
F	F	4

module fails, then the state of the system changes to (FO). So, a Markov process represents possible chains of events which occur within a system. In the case of dependability evaluation, these events are failures or repairs.

Each edge of a state transition diagram carries a label, reflecting the rate at which the state transitions occur. Depending on the modeling goals, this can be failure rate or repair rate.

We illustrate the concept in the example of a simple single-component system.

3.3.4.1 Single-Component System

A single-component system has only two states: operational (state 1) and failed (state 2). If no repairs are allowed, there is a single, nonreversible transition between the states, with the label λ corresponding to the failure rate of the component (see Fig. 3.7). We follow the convention of [9], in which self-transitions are not shown. In some textbooks, self-transitions are represented as loops [7, 14].

If repair is allowed, then a transition between the failed and the operational states is possible, with the label μ corresponding to the repair rate of the component (see Fig. 3.8).

Next, suppose that we would like to distinguish between a failed-safe and a failed-unsafe state, as required in safety evaluation. Let the state 2 be a failed-safe and the state 3 be a fail-unsafe state (see Fig. 3.9). The transition between the states 1 and 2 depends on the component failure rate λ and the fault-detection coverage C. Recall that the fault-detection coverage is the probability that, if a fault exists, the system

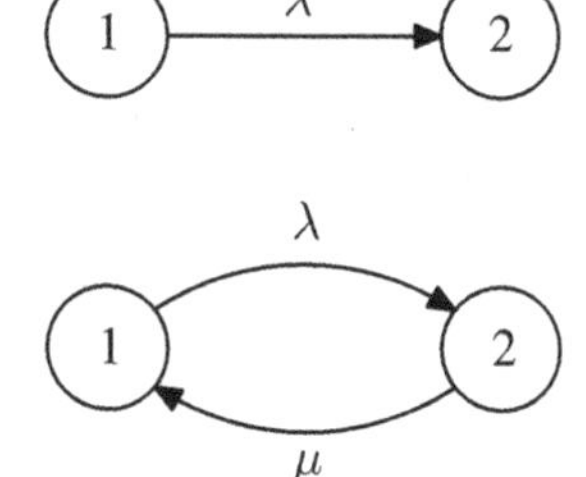

Fig. 3.7 Markov chain of a single-component system

Fig. 3.8 Markov chain of a single-component system with repair

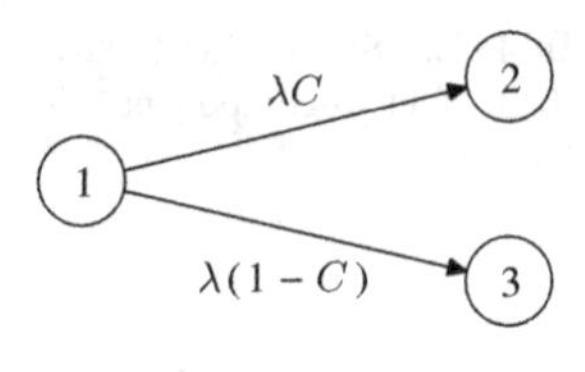

Fig. 3.9 Markov chain of a single-component system for safety evaluation

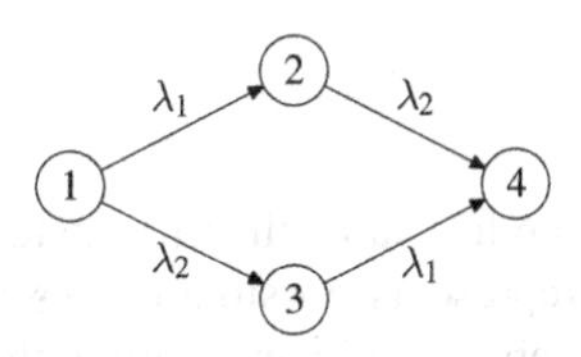

Fig. 3.10 Markov chain of a two-component system

detects it successfully. It is necessary for the system to detect a fault in order to take the corresponding actions to fail in a safe manner.

Similarly, the transition between the states 1 and 3 depends on the failure rate λ and the probability that a fault is *not* detected, i.e. $1 - C$.

3.3.4.2 Two-Component System

A two-component system has four possible states, enumerated in Table 3.2. The changes of states are illustrated by the state transition diagram shown in Fig. 3.10. The failure rates λ_1 and λ_2 of components 1 and 2, respectively, indicate the rates at which the transitions are made between the states. The two components are assumed to be independent and nonrepairable.

If the components are in a serial configuration, then any component failure causes system failure. So, only state 1 is an operational state. States 2, 3, and 4 are failed states. If the components are in a parallel configuration, then both components must fail for the system to fail. Therefore, states 1, 2, and 3 are operational states, whereas state 4 is a failed state.

Next, assume that the components are in parallel and they can be repaired as long as the system does not fail. This means that failed components can be repaired in states 2 and 3, as shown in Fig. 3.11. The repair rates of components 1 and 2 are μ_1 and μ_2, respectively.

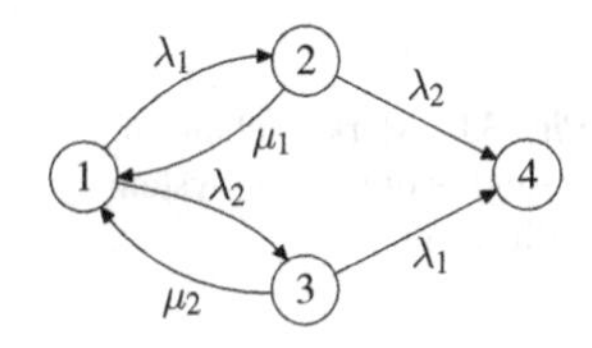

Fig. 3.11 Markov chain of a two-component system with repair

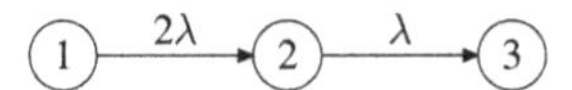

Fig. 3.12 Simplified Markov chain of a two-component system

Table 3.3 States of the simplified Markov chain of a two-component system

Component		State
1	2	Number
O	O	1
O	F	2
F	O	2
F	F	3

3.3.4.3 State Transition Diagram Simplification

It is often possible to reduce the size of a state transition diagram without sacrificing the accuracy of analysis. For example, if the components of the two-component system represented by the state transition diagram in Fig. 3.10 have identical failure rates $\lambda_1 = \lambda_2 = \lambda$, then it is not necessary to distinguish between states 2 and 3. Both states represent a condition where one component is operational and one is failed. So we can merge these two states into one as shown in Fig. 3.12. The assignments of the states in the simplified Markov chain are given in Table 3.3.

Since the failures of components are assumed to be independent events, the transition rate from states 1 to 2 is the sum of the transition rates from states 1 to 2 and 3 in Fig. 3.10, i. e. 2λ.

Example 3.12. Draw a Markov chain for the three-component system whose RBD is shown in Fig. 3.5. Assume that a failed processor can be repaired with the repair rate μ as long as the system has not failed. The failure rate of the processors 1 and 2 is λ_p and the failure rate of the memory is λ_m.

The resulting Markov chain is shown in Fig. 3.13. The description of its states is given in Table 3.4. P_1 and P_2 stand for the processors and M stands for the memory.

3.4 Dependability Evaluation

In this section, we describe how reliability block diagrams and Markov processes can be used for evaluating system dependability.

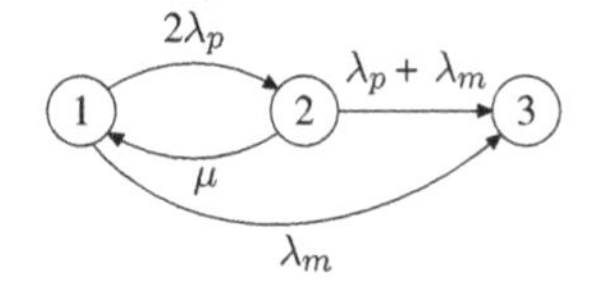

Fig. 3.13 Markov chain of the three-component system whose RBD is shown in Fig. 3.5

Table 3.4 States of the Markov chain in Fig. 3.13

Component			State
P_1	P_2	M	Number
O	O	O	1
O	O	F	3
O	F	O	2
O	F	F	3
F	O	O	2
F	O	F	3
F	F	O	3
F	F	F	3

3.4.1 Reliability Evaluation Using Reliability Block Diagrams

To compute the reliability of a system represented by an RBD, we first partition the system into serial and parallel subsystems. Then, we compute the reliabilities of these subsystems. Finally, the overall solution is composed from the reliabilities of the subsystems.

In a serial system, all components should be operational for a system to function correctly. Hence, by rule (3.5):

$$R_{serial}(t) = \prod_{i=1}^{n} R_i(t), \tag{3.19}$$

where $R_i(t)$ is the reliability of the ith component, for $i \in \{1, 2, \ldots, n\}$.

In a parallel system, only one of the components is required for a system to be operational. So, the unreliability of a parallel system is equal to the probability that all n elements fail:

$$Q_{parallel}(t) = \prod_{i=1}^{n} Q_i(t) = \prod_{i=1}^{n}(1 - R_i(t)). \tag{3.20}$$

Hence, by rule 1:

$$R_{parallel}(t) = 1 - Q_{parallel}(t) = 1 - \prod_{i=1}^{n}(1 - R_i(t)).$$

We can summarize the above as:

$$R(t) = \begin{cases} \prod_{i=1}^{n} R_i(t) & \text{for a series structure,} \\ 1 - \prod_{i=1}^{n}(1 - R_i(t)) & \text{for a parallel structure.} \end{cases} \tag{3.21}$$

Designing a reliable serial system with many components is difficult. For example, if a serial system with 1,000 components is to be build and each of the components has the reliability 0.999, the overall system reliability is only $0.999^{1000} = 0.368$.

On the other hand, a parallel system can be made reliable in spite of the unreliability of its components. For example, a parallel system of four identical components with the component reliability 0.80 has the system reliability $1 - (1 - 0.80)^4 = 0.9984$. Clearly, however, the cost of the parallelism can be high.

Example 3.13. Compute the reliability of the system from Example 3.11.

From the RBD in Fig. 3.6, we can derive the following expression for system reliability:

$$R_{\text{system}}(t) = 1 - (1 - R_A(t)R_B(t))(1 - R_C(t)(1 - (1 - R_D(t))(1 - R_E(t)))),$$

where $R_A(t)$, $R_B(t)$, $R_C(t)$, $R_D(t)$ and $R_E(t)$ are reliabilities of the components A, B, C, D and E, respectively.

Example 3.14. A system consists of three modules: M_1, M_2 and M_3. After analyzing the system, the following reliability expression was derived from its RBD:

$$R_{\text{system}}(t) = R_1(t)R_3(t) + R_2(t)R_3(t) - R_1(t)R_2(t)R_3(t)$$

where $R_i(t)$ is the reliability of the module i, for $i \in \{1, 2, 3\}$. Draw the reliability block diagram of this system.

We can rewrite the reliability expression as:

$$R_{\text{system}}(t) = R_3(t)(R_1(t)+R_2(t)-R_1(t)R_2(t))) = R_3(1-(1-R_1(t))(1-R_2(t))).$$

From the above, we can infer the RBD shown in Fig. 3.14.

3.4.2 Dependability Evaluation Using Markov Processes

In this section, we show how Markov processes can be used to evaluate system dependability. Continuous time markov chains are the most important class of Markov processes for dependability evaluation, so we focus the presentation on this model.

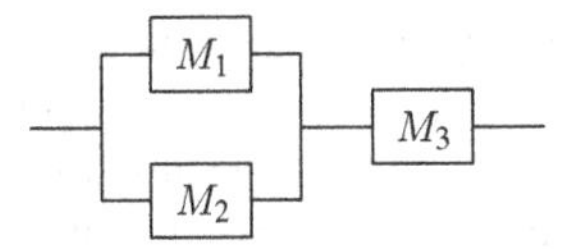

Fig. 3.14 An RBD of the system from Example 3.14

The aim of Markov process analysis is to calculate $P_i(t)$, the probability that the system is in state i at time t. Once this is known, the system reliability, availability, or safety can be computed as the sum of the probabilities of all the operational states.

Suppose that state 1 is the state in which all the components are operational. Assuming that at $t = 0$ the system is in state 1, we get

$$P_1(0) = 1.$$

Since at any time the system can be only in one state, $P_i(0) = 0, \forall i \neq 1$, and we have

$$\sum_{i \in O \cup F} P_i(t) = 1, \tag{3.22}$$

where the sum is over all possible operational (O) and failed (F) states.

To determine the probabilities $P_i(t)$, we derive a set of differential equations, one for each state i of the system. These equations are called *state transition equations*, because they allow the probabilities $P_i(t)$ to be determined in terms of the rates (failure or repair) at which transitions are made from one state to another. State transition equations are usually presented in a matrix form. The matrix M whose entry m_{ij} is the rate of transition between the states i and j is called the *transition matrix*. We use the first index i for the columns of the matrix and the second index j for the rows, i.e. M has the following structure

$$\mathbf{M} = \begin{bmatrix} m_{11} & m_{21} & \dots & m_{k1} \\ m_{12} & m_{22} & \dots & m_{k1} \\ & & \dots & \\ m_{1k} & m_{2k} & \dots & m_{kk} \end{bmatrix},$$

where k is the number of states in a Markov chain representing the system. The entries m_{ii} corresponding to self-transitions are computed as $-\sum m_{ij}$, for all $i, j \in \{1, 2, \dots k\}$ such that $i \neq j$. Thus, the entries in each column of the transition matrix sum up to 0. A positive sign in an ijth entry indicates that the transition originates in the ith state. A negative sign in an ijth entry indicates that the transition terminates in the ith state.

For example, the transition matrix for the Markov chain of a single-component system in Fig. 3.7 is

$$\mathbf{M} = \begin{bmatrix} m_{11} & m_{21} \\ m_{12} & m_{22} \end{bmatrix} = \begin{bmatrix} -\lambda & 0 \\ \lambda & 0 \end{bmatrix}. \tag{3.23}$$

The rate of transition between states 1 and 2 is λ, therefore the $m_{12} = \lambda$ and $m_{11} = -\lambda$. The rate of transition between states 2 and 1 is 0, so $m_{21} = 0$ and $m_{22} = 0$.

In reliability evaluation, once a system fails, the failed state cannot be left. Therefore, we can use a transition matrix to distinguish between the operational and failed states. Each failed state i has a zero diagonal element m_{ii}. This is not the case, however, when availability or safety is computed, as we will see in the examples below.

The transition matrix for the Markov chain in Fig. 3.8, which incorporates repair, is

$$\mathbf{M} = \begin{bmatrix} -\lambda & \mu \\ \lambda & -\mu \end{bmatrix}. \tag{3.24}$$

The transition matrix for the Markov chain in Fig. 3.9, is size 3×3 since, for safety evaluation, we assume that the system can take three different states: operational, failed-safe failed-unsafe:

$$\mathbf{M} = \begin{bmatrix} -\lambda & 0 & 0 \\ \lambda C & 0 & 0 \\ \lambda(1-C) & 0 & 0 \end{bmatrix}. \tag{3.25}$$

The transition matrix for the simplified state transition diagram of the two-component system, shown in Fig. 3.12 is

$$\mathbf{M} = \begin{bmatrix} -2\lambda & 0 & 0 \\ 2\lambda & -\lambda & 0 \\ 0 & \lambda & 0 \end{bmatrix}. \tag{3.26}$$

Using state transition matrices, state transition equations are derived as follows. Let $\mathbf{P}(t)$ be a vector whose ith element is the probability $P_i(t)$ that the system is in state i at time t. Then the matrix representation of a system of state transition equations is given by

$$\frac{\mathrm{d}}{\mathrm{d}t}\mathbf{P}(t) = \mathbf{M} \times \mathbf{P}(t). \tag{3.27}$$

Once the system of equations is solved and the probabilities $P_i(t)$ are known, system reliability, availability, or safety can be computed as a sum of probabilities taken over all the operational states O:

$$R(t) = \sum_{i \in O} P_i(t), \tag{3.28}$$

or, alternatively:

$$R(t) = 1 - \sum_{i \in F} P_i(t),$$

where the sum is taken over all failed states F.

We illustrate the computation process on a number of simple examples.

3.4.2.1 Reliability Evaluation

We start with the case of independent components and then proceed with the case of dependent components.

Independent Components
Let us first compute the reliability of a parallel system consisting of two independent components which we have considered before (Fig. 3.10). Applying (3.27) to the matrix (3.26) we get

$$\frac{d}{dt}\begin{bmatrix} P_1(t) \\ P_2(t) \\ P_3(t) \end{bmatrix} = \begin{bmatrix} -2\lambda & 0 & 0 \\ 2\lambda & -\lambda & 0 \\ 0 & \lambda & 0 \end{bmatrix} \times \begin{bmatrix} P_1(t) \\ P_2(t) \\ P_3(t) \end{bmatrix}.$$

The above matrix form represents the following system of state transition equations

$$\begin{cases} \frac{d}{dt}P_1(t) = -2\lambda P_1(t) \\ \frac{d}{dt}P_2(t) = 2\lambda P_1(t) - \lambda P_2(t) \\ \frac{d}{dt}P_3(t) = \lambda P_2(t) \end{cases}$$

By solving this system of equations, we get

$$\begin{aligned} P_1(t) &= e^{-2\lambda t} \\ P_2(t) &= 2e^{-\lambda t} - 2e^{-2\lambda t} \\ P_3(t) &= 1 - 2e^{-\lambda t} + e^{-2\lambda t} \end{aligned}$$

Since the probabilities $P_i(t)$ are known, we can now calculate the reliability of the system. For a serial configuration, the failure of the first component causes the system failure. Therefore, the reliability of the system equals the probability $P_1(t)$:

$$R_{\text{serial}}(t) = e^{-2\lambda t}. \tag{3.29}$$

Since the reliability of a single component is $R(t) = e^{-\lambda t}$, we get $R_{\text{serial}}(t) = R^2(t)$ which agrees with the expression (3.19) obtained from RBDs.

For a parallel configuration, both components should fail for the system to fail. Therefore, the reliability of the system is the sum of probabilities $P_1(t)$ and $P_2(t)$:

$$R_{\text{parallel}}(t) = 2e^{-\lambda t} - e^{-2\lambda t}. \tag{3.30}$$

We can rewrite the above as $R_{\text{serial}}(t) = 2R(t) - R^2(t) = 1 - (1 - R(t))^2$ which agrees with the expression 3.20 obtained from RBDs.

The results of Markov chain and RBD analysis are the same, because in this example we assumed that the component failures are mutually independent. Next, we consider the case of dependent components.

Dependent Components

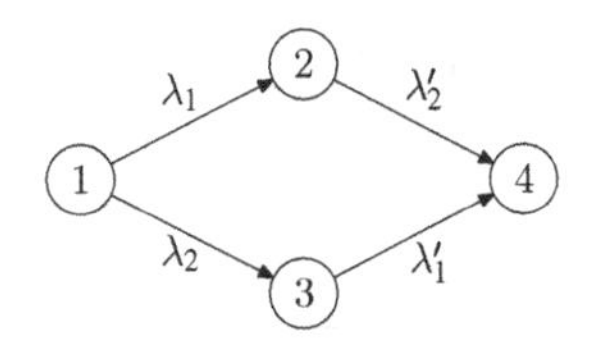

Fig. 3.15 State transition diagram of a two-component parallel system with load sharing

The value of Markov processes becomes evident in situations in which component failure rates depend on the system state. A typical example is load sharing components, which we consider in this section. Another example is standby components, considered in Sect. 3.4.2.2.

The word *load* is used in the broad sense of the stress on a system. This can be an electrical load, a load caused by high temperature, or an information load. In practice, failure rates typically increase as the load increases. Suppose that several components share a load. If one of the component fails, the additional load on the remaining component is likely to increase their failure rate. For example, if one tire in a car blows out, the vertical load on other tires increases. This increases their failure rate.

As an example, consider the Markov chain of a two-component parallel system shown in Fig. 3.15. The states are defined as previously (Table 3.2). After failure of the first components, the failure rate of the second component increases. The increased failure rates of components 1 and 2 are denoted by λ_1' and λ_2', respectively.

From the Markov chain in Fig. 3.15, we can derive the state transition equations in the matrix form:

$$\frac{d}{dt}\begin{bmatrix} P_1(t) \\ P_2(t) \\ P_3(t) \\ P_4(t) \end{bmatrix} = \begin{bmatrix} -\lambda_1-\lambda_2 & 0 & 0 & 0 \\ \lambda_1 & -\lambda_2' & 0 & 0 \\ \lambda_2 & 0 & -\lambda_1' & 0 \\ 0 & \lambda_2' & \lambda_1' & 0 \end{bmatrix} \times \begin{bmatrix} P_1(t) \\ P_2(t) \\ P_3(t) \\ P_4(t) \end{bmatrix}.$$

By expanding the matrix form, we get the following system of equations:

$$\begin{cases} \frac{d}{dt}P_1(t) = (-\lambda_1-\lambda_2)P_1(t) \\ \frac{d}{dt}P_2(t) = \lambda_1 P_1(t) - \lambda_2' P_2(t) \\ \frac{d}{dt}P_3(t) = \lambda_2 P_1(t) - \lambda_1' P_3(t) \\ \frac{d}{dt}P_4(t) = \lambda_2' P_2(t) + \lambda_1' P_3(t). \end{cases}$$

The solution of this system of equations is given by:

$$P_1(t) = e^{(-\lambda_1-\lambda_2)t}$$

$$P_2(t) = \frac{\lambda_1}{\lambda_1+\lambda_2-\lambda_2'}e^{\lambda_2' t} - \frac{\lambda_1}{\lambda_1+\lambda_2-\lambda_2'}e^{(-\lambda_1-\lambda_2)t}$$

$$P_3(t) = \frac{\lambda_2}{\lambda_1+\lambda_2-\lambda_1'}e^{\lambda_1' t} - \frac{\lambda_2}{\lambda_1+\lambda_2-\lambda_1'}e^{(-\lambda_1-\lambda_2)t}$$

$$P_4(t) = 1 - e^{(-\lambda_1-\lambda_2)t} - \frac{\lambda_1}{\lambda_1+\lambda_2-\lambda_2'}e^{\lambda_2' t} + \frac{\lambda_1}{\lambda_1+\lambda_2-\lambda_2'}e^{(-\lambda_1-\lambda_2)t} - \frac{\lambda_2}{\lambda_1+\lambda_2-\lambda_1'}e^{\lambda_1' t} + \frac{\lambda_2}{\lambda_1+\lambda_2-\lambda_1'}e^{(-\lambda_1-\lambda_2)t}.$$

Finally, since $P_4(t)$ is the failed state, the reliability of the system is equal to $1-P_4(t)$:

$$R_{\text{parallel}}(t) = e^{(-\lambda_1-\lambda_2)t} + \frac{\lambda_1}{\lambda_1+\lambda_2-\lambda_2'}e^{\lambda_2' t} - \frac{\lambda_1}{\lambda_1+\lambda_2-\lambda_2'}e^{(-\lambda_1-\lambda_2)t} + \frac{\lambda_2}{\lambda_1+\lambda_2-\lambda_1'}e^{\lambda_1' t} - \frac{\lambda_2}{\lambda_1+\lambda_2-\lambda_1'}e^{(-\lambda_1-\lambda_2)t}. \quad (3.31)$$

The reader can verify that if $\lambda_1' = \lambda_1$ and $\lambda_2' = \lambda_2$, the above equation is equal to the Eq. (3.30).

The effect of the increased load can be illustrated as follows. Suppose that the two components are identical, i.e. $\lambda_1 = \lambda_2 = \lambda$ and $\lambda_1' = \lambda_2' = \lambda'$. Then, the Eq. (3.31) reduces to

$$R_{\text{parallel}}(t) = \frac{2\lambda}{2\lambda-\lambda'}e^{-\lambda' t} - \frac{\lambda'}{2\lambda-\lambda'}e^{-2\lambda t}.$$

Figure 3.16 shows the reliability of a two-component parallel system with load sharing for different values of λ'. The reliability $e^{-\lambda t}$ of a single-component system is also plotted for a comparison. The case of $\lambda' = \lambda$ corresponds to the independent components, so the reliability is given by the Eq. (3.27). The case of $\lambda' = \infty$ is the case of total dependency. This means that the failure of one component would bring an immediate failure of the other component. So, in this case, the reliability is equal to the reliability of a serial system with two components given by the Eq. (3.21). We can see that, the more the value of λ' exceeds the value of λ, the closer the reliability of this system approaches the reliability of a serial system with two components.

3.4.2.2 Availability Evaluation

The main difference between reliability and availability evaluation is in the treatment of repair. In reliability evaluation, components can be repaired until the system has not failed. In availability evaluation, the components can also be repaired *after* system failure.

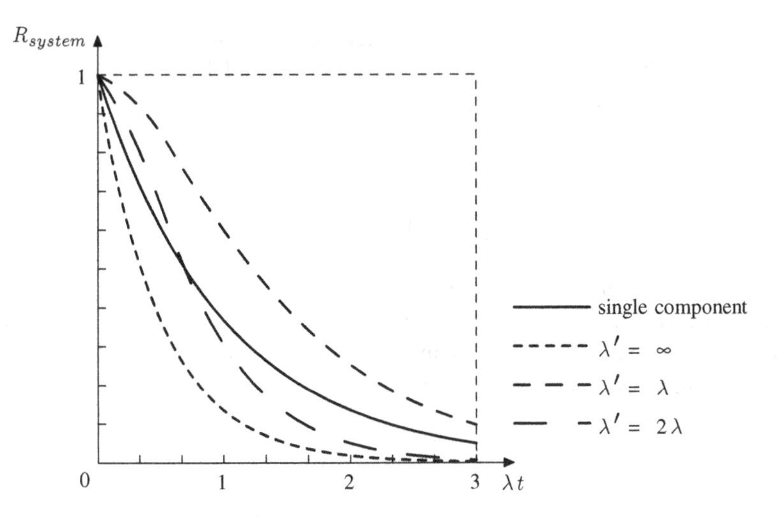

Fig. 3.16 Reliability of a two-component parallel system with load sharing

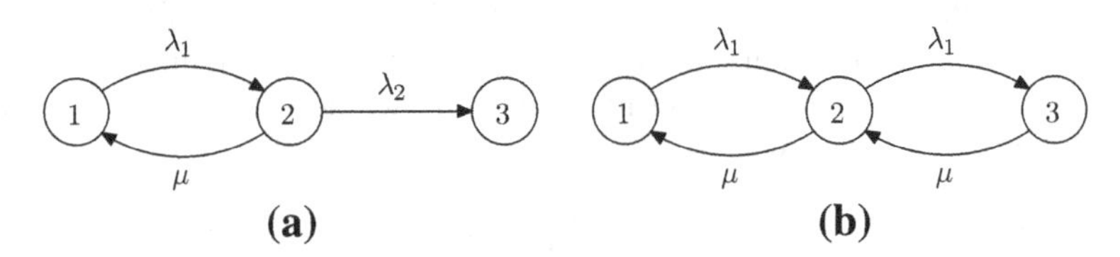

Fig. 3.17 State transition diagrams for a standby two-component system **a** for reliability evaluation, **b** for availability evaluation

Table 3.5 Markov states of a simplified state transition diagram of a two-component parallel system incorporating repair

Component		State
Primary	Spare	Number
O	O	1
F	O	2
F	F	3

This difference is best illustrated in the example of a standby system with two components, one primary and one spare. The spare component is held in reserve and only brought into operation when the primary component fails. We assume that there is a perfect fault-detection unit which detects failures in the primary component and replaces it by the spare. We also assume that the spare cannot fail while it is in standby mode.

The Markov chains of the standby system for reliability and availability evaluation are shown in Fig. 3.17a, b, respectively. The states are defined according to Table 3.5. When the primary component fails, there is a transition between states 1 and 2. If the system is in state 2 and the spare component fails, there is a transition to state 3. Since we assumed that the spare cannot fail while in the standby mode, the combination (O, F) cannot occur. States 1 and 2 are operational states. State 3 is the failed state.

Suppose the primary component can be repaired at a rate μ. For reliability evaluation, this implies that a transition between states 2 and 1 is possible. The corresponding transition matrix is given by

$$\mathbf{M} = \begin{bmatrix} -\lambda_1 & \mu & 0 \\ \lambda_1 & -\lambda_2 - \mu & 0 \\ 0 & \lambda_2 & 0 \end{bmatrix}.$$

For availability evaluation, we should be able to repair from a failed state as well. This adds a transition between states 3 and 2. We assume that the repair rates for the primary and the spare units are the same and that the spare unit is repaired first. The corresponding transition matrix is given by

$$\mathbf{M} = \begin{bmatrix} -\lambda_1 & \mu & 0 \\ \lambda_1 & -\lambda_2 - \mu & \mu \\ 0 & \lambda_2 & -\mu \end{bmatrix}. \tag{3.32}$$

We can see that, in the matrix for availability calculations, none of the diagonal elements is zero. This is because the system is able to recover from the failed state.

By solving the system of state transition equations, we can get $P_i(t)$ and compute the availability of the system as:

$$A(t) = \sum_{i \in O} P_i(t), \tag{3.33}$$

where the sum is taken over all the failed states O, or

$$A(t) = 1 - \sum_{i \in F} P_i(t), \tag{3.34}$$

where the sum is taken over all the failed states F.

As we mentioned in Sect. 2.2.2, often the steady-state availability rather than the time-dependent availability is of interest. The steady-state availability can be computed in a simpler way. We note that, as time approaches infinity, the derivative on the right-hand side of the Eq. (3.27) vanishes and we get a time-independent relationship

$$\mathbf{M} \times \mathbf{P}(\infty) = 0. \tag{3.35}$$

In our example, for matrix (3.32) this represents a system of equations

$$\begin{cases} -\lambda_1 P_1(\infty) + \mu P_2(\infty) = 0 \\ \lambda_1 P_1(\infty) - (\lambda_2 + \mu) P_2(\infty) + \mu P_3(\infty) = 0 \\ \lambda_2 P_2(\infty) - \mu P_3(\infty) = 0 \end{cases}$$

Since these three equations are linearly dependent, they are not sufficient to solve for $P(\infty)$. The required piece of additional information is the condition (3.22) that the sum of all probabilities is one:

$$\sum_i P_i(\infty) = 1. \tag{3.36}$$

If we assume $\lambda_1 = \lambda_2 = \lambda$, then we get

$$P_1(\infty) = \left[1 + \frac{\lambda}{\mu} + \left(\frac{\lambda}{\mu}\right)^2\right]^{-1}$$

$$P_2(\infty) = \left[1 + \frac{\lambda}{\mu} + \left(\frac{\lambda}{\mu}\right)^2\right]^{-1} \frac{\lambda}{\mu}$$

$$P_3(\infty) = \left[1 + \frac{\lambda}{\mu} + \left(\frac{\lambda}{\mu}\right)^2\right]^{-1} \left(\frac{\lambda}{\mu}\right)^2.$$

The steady-state availability can be found by setting $t = \infty$ in (3.34)

$$A(\infty) = 1 - \left[1 + \frac{\lambda}{\mu} + \left(\frac{\lambda}{\mu}\right)^2\right]^{-1} \left(\frac{\lambda}{\mu}\right)^2.$$

If we further assume that $\lambda/\mu << 1$, we obtain

$$A(\infty) \approx 1 - \left(\frac{\lambda}{\mu}\right)^2.$$

3.4.2.3 Safety Evaluation

The main difference between safety and reliability evaluation is that, for safety evaluation, the failed state is split into failed-safe and failed-unsafe ones. Once the Markov chain for a system is constructed, the state transition equations are obtained using the same procedure as for reliability evaluation. Once the system of equations is solved, safety is computed as the sum of probabilities taken over all operational and failed-safe states.

As an example, consider the single component system shown in Fig. 3.9. Its state transition matrix is given by (3.25). So, the state transition equations for $P_i(t)$ are given by

$$\frac{d}{dt}\begin{bmatrix} P_1(t) \\ P_2(t) \\ P_3(t) \end{bmatrix} = \begin{bmatrix} -\lambda & 0 & 0 \\ \lambda C & 0 & 0 \\ \lambda(1-C) & 0 & 0 \end{bmatrix} \times \begin{bmatrix} P_1(t) \\ P_2(t) \\ P_3(t) \end{bmatrix}.$$

The solution of this system of equations is

$$P_1(t) = e^{-\lambda t}$$
$$P_2(t) = C - Ce^{-\lambda t}$$
$$P_3(t) = (1 - C) - (1 - C)e^{-\lambda t}.$$

The safety of the system is the sum of probabilities of being in the operational or fail-safe states:

$$S(t) = P_1(t) + P_2(t) = C + (1 - C)e^{-\lambda t}.$$

At time $t = 0$, the safety of the system is 1. As time approaches infinity, the safety approaches the fault-detection coverage C. So, if $C = 1$, the system has a perfect safety.

3.5 Summary

In this chapter, we have considered common dependability measures, such as failure rate, mean time to failure, mean time to repair, mean time between failures, and fault coverage. We have explored combinatorial dependability models such as reliability block diagrams, fault trees, and reliability graphs. We have also studied stochastic dependability models such as Markov chains, which enable the analysis of more complex scenarios. Finally, we have discussed how these models can be used for evaluating system reliability, availability, and safety.

Problems

In all problems, we assume that failures and repairs of components are independent events (unless it is specified otherwise) and failure and repair rates are constant.

3.1. A heart pacemaker has a failure rate of $\lambda = 0.121 \times 10^{-8}$ per hour.

1. What is the probability that it fails during the first five years of operation?
2. What is its MTTF?

3.2. Compute the reliability of the majority voter shown in Fig. 4.8 for 1 year of service assuming that the two-input AND gates and the three-input OR gate have the failure rates 0.324×10^{-7} and 0.381×10^{-7} per hour, respectively.

3.3. Compute the MTTF of the full adder shown in Fig. 5.11 assuming that the two-input AND and OR gates have the failure rate 0.324×10^{-7} and the two-input XOR gate has the failure rate 0.336×10^{-7} per hour, respectively.

3.4. A manufacturer of refrigerators determines that their product has MTTF = 6×10^5 h in normal use. For how long should the warranty be set if no more than 5 % of the refrigerators are to be returned for repair?

3.5. A modem manufacturer estimates that the reliability of their product is 72.8 % during the first three years of operation in normal use.

1. How many modems will need a repair during the first year of operation?
2. What is the MTTF of the modems?

3.6. A two-year guarantee is given on a TV-set, based on the assumption that no more than 3 % of the items will be returned for repair. What is the maximum failure rate that can be tolerated?

3.7. A DVD-player manufacturer determines that the average DVD set is used 930 hr/year. A two-year warranty is offered on the DVD set having an MTTF of 25×10^3 hours. Assuming that a DVD cannot fail while it is not in use, what fraction of DVD sets will fail during the warranty period?

3.8. Suppose that a jet engine has a constant failure rate of $\lambda = 0.214 \times 10^{-7}$ per hour. What is the probability that more than two engines on a four-engine aircraft will fail during an eight-hour flight?

3.9. A nonredundant system with 50 components has a design life reliability of 0.95. The system is re-designed so that it has only 35 components. Estimate the design life reliability of the modified system. Assume that all the components have the same failure rate.

3.10. At the end of the year of service, the reliability of a component is 0.96.

1. What is the failure rate of the component?
2. If two components are connected in parallel, what will be the reliability of the resulting system during the first year of operation?

3.11. A lamp has three bulbs. The failure rate of a bulb is $\lambda = 0.182 \times 10^{-3}$ per year. What is the probability that more than one bulb fails during the first 6 months of operation?

3.12. Suppose that in Problem 3.11 bulbs are load sharing components. Initially, the failure rate of the bulb i is λ_i, for $i \in \{1, 2, 3\}$. When the first bulb fails, the failure rates of the remaining bulbs increase to λ_i'. When the second bulb fails, the failure rates of the remaining bulb becomes λ_i''. Draw a Markov chain for reliability evaluation of this system.

3.13. Suppose a component has a failure rate of $\lambda = 0.184 \times 10^{-3}$ per hour. How many components should be placed in parallel if the system is to run for 800 hours with a failure probability of no more than 1 %?

3.14. A system has MTTF = 12×10^3 h. An engineer is to set the design life time, so that the end-of-life reliability is 95 %.

1. Determine the design life time.
2. If two systems are placed in parallel, to what value may the design life time be increased without decreasing the end-of-life reliability?

3.15. A printer has a failure rate of $\lambda = 0.286 \times 10^{-4}$ per hour and an MTTR = 72 hours in normal use.

1. What is its steady-state availability?
2. If MTTR is increased to 120 h, what failure rate can be tolerated without decreasing the availability of the printer?

3.16. A fax machine has a failure rate of 0.102×10^{-2} per hour in normal use. What repair rate should be maintained to achieve a steady-state availability of 97 %?

3.17. Suppose that a standby system with two components is required to have the steady-state availability 0.91. What is the maximum acceptable value of the failure-to-repair ratio λ/μ?

3.18. A computer system is designed to have a failure rate of one fault in 5 years in normal use. The system has no fault tolerance capabilities, so it fails upon occurrence of the first fault.

1. What is the MTTF of such a system?
2. What is the probability that the system will fail during its first year of operation?
3. The usual warranty for the system is 2 years. The vendor wishes to offer an additional insurance against failures for the first 5 years of operation at extra cost. The vendor wants to charge $20 for each 1 % drop in reliability to offer such an insurance. How much should the vendor charge for such an insurance?

3.19. A system with four modules: A, B, C, D and, E is connected, so that it operates correctly if (1) modules A or D operate correctly, and (2) modules C and D operate correctly, or module E operates correctly.

1. Draw an RBD of the system.
2. Write an expression for the reliability of the system.
3. Compute the reliability of the system for a 3-year mission, assuming that modules have the following failure rates per hour: $\lambda_A = 0.318\times10^{-6}$, $\lambda_B = 0.256\times10^{-6}$, $\lambda_C = 0.431\times10^{-6}$, $\lambda_D = 0.417\times10^{-6}$, $\lambda_E = 0.372\times10^{-6}$.

3.20. Suppose that the reliability of a system consisting of four blocks, two of which are identical, is given by the following equation:

$$R_{\text{system}}(t) = R_1(t)R_2(t)R_3(t) + R_1^2(t) - R_1^2(t)R_2(t)R_3(t) \tag{3.37}$$

Draw the reliability block diagram representing the system.

3.21. Write an expression for the reliability of the system shown by the RBD in Fig. 3.18. Assume that $R_i(t)$ is the reliability of the module M_i, for $i \in \{1, 2, 3, 4\}$.

3.22. Your company produces a system which consists of a memory and a processor. To increase the system's reliability, the memory is duplicated and the processor is triplicated (see RBD in Fig. 3.19).

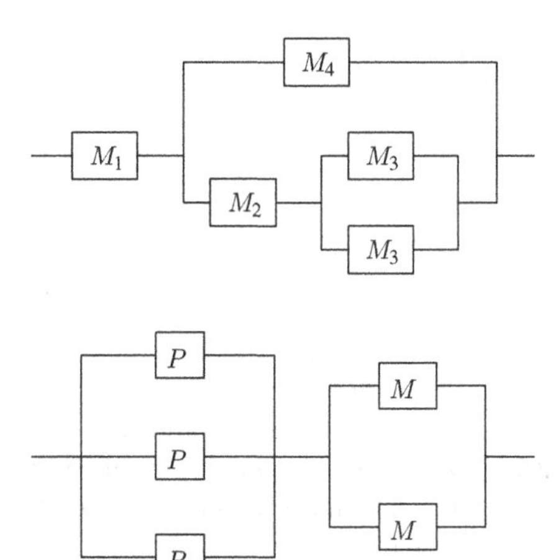

Fig. 3.18 RBD of the system for Problem 3.21

Fig. 3.19 RBD of the system for Problems 3.22 and 3.26

The reliabilities of the memory and the processor for the first year of operation are $R_M(1\text{year}) = 0.95$ and $R_P(1\text{year}) = 0.9$, and their costs are 50 and 30$, respectively.
Your boss decides that the system is too expensive. She would like you to reduce its cost by using processors with a higher reliability. There are three manufacturers producing processors which are suitable for your system. They offer processors with the following reliabilities and costs:

1. $R_{P_1}(1\text{year}) = 0.9683$ for 40$
2. $R_{P_2}(1\text{year}) = 0.9812$ for 45$
3. $R_{P_3}(1\text{year}) = 0.9989$ for 50$

Your task is to estimate which processor minimizes the cost of the system without decreasing its reliability compared to the original system. You are allowed to combine processors from different manufacturers.
Draw an RBD of the new system.

3.23. Your company produces a system which consists of two components, A and B, placed in series. The reliabilities of the components for the first year of operation are $R_A(1\text{year}) = 0.99$ and $R_B(1\text{year}) = 0.85$. Their cost is the same.
The warranty for this system is 1 year. Your boss decides that too many items are returned for repair during the warranty period. She gives you the task of improving the reliability of the system, so that no more than 2 % of the items are returned. This should be done by adding no more than two redundant components to the system.
Draw an RBD for the system which meets your boss's requirements.

3.24. How many states have a nonsimplified Markov chain for a system consisting of n components? Assume that each component has two states: operational and failed.

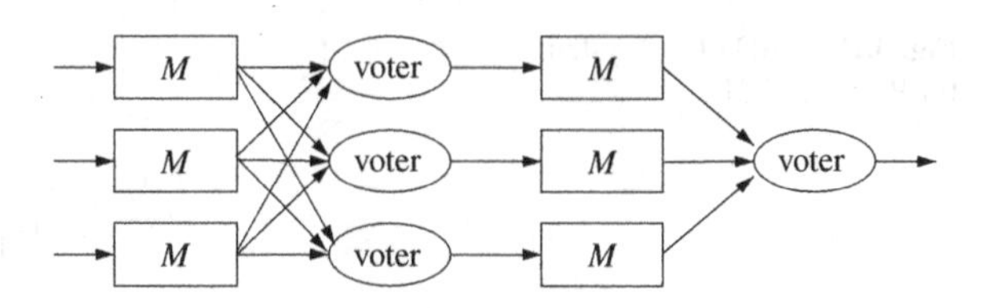

Fig. 3.20 Diagram of the system for Problem 3.28

3.25. Draw a Markov chain for reliability evaluation of the system shown in Fig. 3.5. The failure rate of the processors 1 and 2 is λ_p. The failure rates of the memory is λ_m. No repairs are allowed.

3.26. Draw a Markov chain for reliability evaluation of the five-component system shown by the RBD in Fig. 3.19. The system consists of a triplicated processor, P, and a duplicated memory, M. The failure rates of the processor and the memory are λ_p and λ_m, respectively. Assume that no repairs are allowed.

3.27. Suppose that a system was modeled using the Markov chain model and the following transition matrix was obtained from the resulting chain:

$$\begin{bmatrix} -(5\lambda+\lambda_v) & \mu & 0 & 0 \\ 5\lambda & -(\mu+4\lambda+\lambda_v) & 2\mu & 0 \\ 0 & 4\lambda & -(2\mu+3\lambda+\lambda_v) & 0 \\ \lambda_v & \lambda_v & 3\lambda+\lambda_v & 0 \end{bmatrix}$$

Draw the Markov chain corresponding to this transition matrix. Was the Markov chain intended for reliability or availability evaluation?

3.28. Draw a Markov chain for reliability evaluation of the system shown in Fig. 3.20. The failure rates of modules and voters are λ and λ_v, respectively. No repairs are allowed.

3.29. Draw a Markov chain for availability evaluation of the system shown in Fig. 3.5. The failure and repair rates of the processors 1 and 2 is λ_p and μ_p, respectively. The failure and repair rates of the memory are λ_m and μ_m, respectively.

3.30. Consider the system shown in Fig. 3.21. This system consists of nine identical modules M connected into three "triples" and four threshold voters. The system works as follows. For each triple, the result of the voter v_1 is compared to the outputs of individual modules M to detect disagreement. If a disagreement occurs, the corresponding switch s_1 opens and disconnects the faulty module. After the removal of the 1st module from a triple, the voter v_1 works a comparator. When the 2nd module in the triple fails, v_1 sends a signal to the switch s_2 which opens s_2 and removes v_1 from the voting in v_2.

After the removal of the 1st triple, the voter v_2 works as a comparator. After the removal of the 2nd triple, the voter v_2 just passes the signal through, so it is sufficient for one triple to work correctly to produce a correct answer.

Construct a Markov chain for reliability evaluation of this system. Assume that the switches and voters are perfect, and the failure rate of each module is λ. No repairs are allowed.

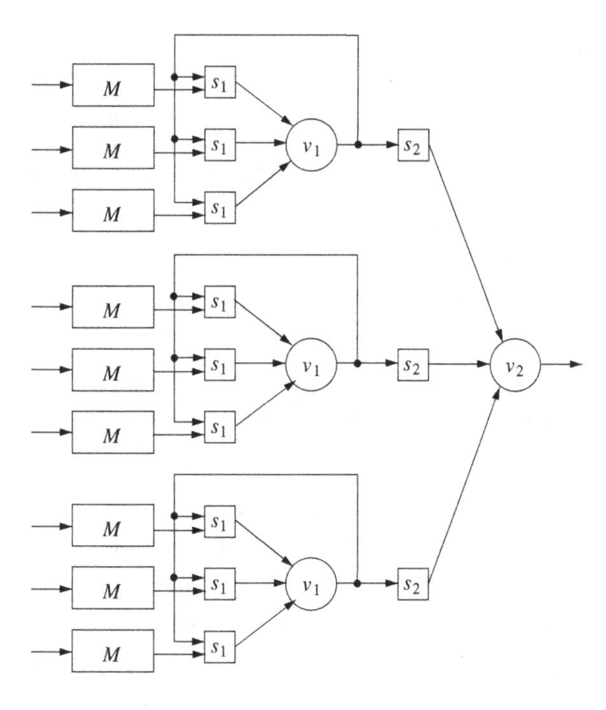

Fig. 3.21 Diagram of the system for Problem 3.30

3.31. You need to buy chains for your car for a 36 h mountain trip in the winter. Local police require that you have chains on all four wheels to drive safely. The shop offers you two choices of sets of chains. Each set consists of four identical chains:

- Set 1: Each chain has the failure rate $\lambda_1 = 0.417 \times 10^{-2}$ per hour. The cost of the set is 20$.
- Set 2: Each chain has the failure rate $\lambda_1 = 0.167 \times 10^{-1}$ per hour. The cost of the set is 10$.

Solve the following tasks:

1. Compute the reliability of your car during the trip, $R_1(t)$, for the case when you buy one Set 1.
2. Draw a Markov chain for reliability evaluation for the case when you buy two Sets 2. Derive and solve the system of equations. Compute the reliability of your car during the trip, $R_2(t)$.
   Based on (1) and (2), which chains do you recommend buying?
3. Compute the lower bound $R_{2lb}(t)$ for the reliability $R_2(t)$ by assuming that spare chains fail at the same failure rate λ_2 as chains in the operation. With this assumption, you can model a car with two sets of cheaper chains as a parallel system with eight components and use RBD to compute $R_{2lb}(t)$. Compare $R_{2lb}(t)$ with $R_2(t)$.

References

1. Crow, L.: Methods for assessing reliability growth potential. In: IEEE Proceedings Annual Reliability and Maintainability Symposium, pp. 484–489 (1984)
2. de Mercado, J., Bowen, N.A.: A method for calculation of network reliability. IEEE Trans Reliab **R-25**, 71–76 (1976)
3. Duane, J.: Learning curve approach to reliability monitoring. IEEE Trans. Aerosp. **2**, 563–566 (1964)
4. Ericson, C.: Fault tree analysis—a history. In: Proceedings of the 17th International Systems Safety Conference (1999)
5. Feller, W.: An Introduction to Probability Theory and Its Applications, 3rd edn. Willey, New York (1968)
6. IEEE Standard 500: IEEE guide to the collection and presentation of electrical, electronic, sensing component, and mechanical equipment reliability data for nuclear-power generating stations (1984)
7. Johnson, B.W.: The Design and Analysis of Fault Tolerant Digital Systems. Addison-Wesley, New York (1989)
8. MIL-HDBK-338: Electronic reliability design handbook. U. S. Department of Defense (1998)
9. Norris, J.R.: Markov Chains. Cambridge University Press, New York (1998)
10. NPRD: Nonelectronic parts reliability data (2011). http://theriac.org/
11. OREDA: Offshore reliability data database (1997). http://www.oreda.com/
12. Rennels, D.: Fault-tolerant computing - concepts and examples. IEEE Trans Comput **C-33**(12), 1116–1129 (1984)
13. RIAC: System Reliability Toolkit. Reliability Information Analysis Center (2005)
14. Shooman, M.L.: Reliability of Computer Systems and Networks: Fault Tolerance, Analysis, and Design. Wiley-Interscience, New York (2001)
15. Siewiorek, D.P., Swarz, R.S.: Reliable Computer Systems Design and Evaluation 3rd ed. A K Peters Ltd., Wellesley (1998)
16. Smith, D.J.: Reliability Engineering. Barnes and Noble Books, New York (1972)
17. Vesely, N.E., R.E.N.: REP and KITT: Computer codes for the automatic evaluation of a fault-tree. Technical report. IN-1349, Idaho Nuclear (1970)

Chapter 4
Hardware Redundancy

"Those parts of the system that you can hit with a hammer (not advised) are called hardware; those program instructions that you can only curse at are called software."

Anonymous

Hardware redundancy is achieved by providing two or more physical copies of a hardware component. For example, a system may contain redundant processors, memories, buses, or power supplies. When other techniques, such as use of more reliable components, manufacturing quality control, test, design simplification, etc., have been exhausted, hardware redundancy may be the only way to improve the dependability of a system. For example, in situations in which equipment cannot be maintained, e.g., communication satellites, redundant components allow uninterrupted operating time to be prolonged.

Hardware redundancy brings a number of penalties: increase in weight, size, power consumption, cost, as well as time to design, fabricate, and test. A number of choices have to be examined to determine the best way to incorporate redundancy into a system. For example, weight increase can be reduced by applying redundancy to the higher level components. Cost increase can be minimized if the expected improvement in dependability reduces the cost of preventive maintenance for the system.

There are three types of hardware redundancy: passive, active, and hybrid. *Passive redundancy* achieves fault tolerance by masking the faults that occur without requiring any action from the system or an operator. *Active redundancy* requires a fault to be detected before it is tolerated. After the detection of the fault, the actions of location, containment and recovery are performed to remove the faulty component from the system. Active techniques require that a system is stopped and reconfigured to tolerate faults. *Hybrid redundancy* combines passive and active approaches. Fault masking is used to prevent generation of erroneous results. Fault detection, location, and recovery are used to replace the faulty component with a spare. Hybrid redundancy enables reconfiguration with no system downtime.

E. Dubrova, *Fault-Tolerant Design*, DOI: 10.1007/978-1-4614-2113-9_4,

In this chapter, we consider a number of different passive, active, and hybrid redundancy configurations and evaluate their effect on system dependability.

4.1 Redundancy Allocation

Originally, redundancy techniques were used for coping with the low reliability of basic hardware components. Designers of early computing systems triplicated low-level components such as gates or flip-flops and used majority voting to correct faults [10]. As the reliability of basic components improved, redundancy was shifted to higher levels. Larger components, such as memories or processor units, became replicated. This decreased the size and probability of failure of voters relative to that of redundant components.

The use of redundancy does not immediately guarantee an improvement in the dependability of a system. The overall increase in complexity caused by redundancy can be quite severe. It may diminish the dependability improvement, unless redundant resources are allocated in a proper way. A careful analysis has to be performed to show that a more dependable system is obtained at the end.

A number of possibilities have to be examined to determine at which level it is best to provide redundancy and which components should be made redundant. To understand the importance of these decisions, consider a serial system consisting of two components with reliabilities R_1 and R_2. If the system reliability $R = R_1 R_2$ does not satisfy the design requirements, the designer may decide to duplicate some of the components. Possible choices of redundant configurations are shown in Fig. 4.1a, b. Assuming the component failures are mutually independent, the corresponding reliabilities of these systems are

$$R_a = (2R_1 - R_1^2)R_2$$
$$R_b = (2R_2 - R_2^2)R_1.$$

Taking the difference of R_b and R_a, we get

$$R_a - R_b = R_1 R_2 (R_2 - R_1).$$

It follows from this expression that the higher reliability is achieved if we duplicate the least reliable component. If $R_1 < R_2$, then configuration Fig. 4.1a is preferable, and vice versa.

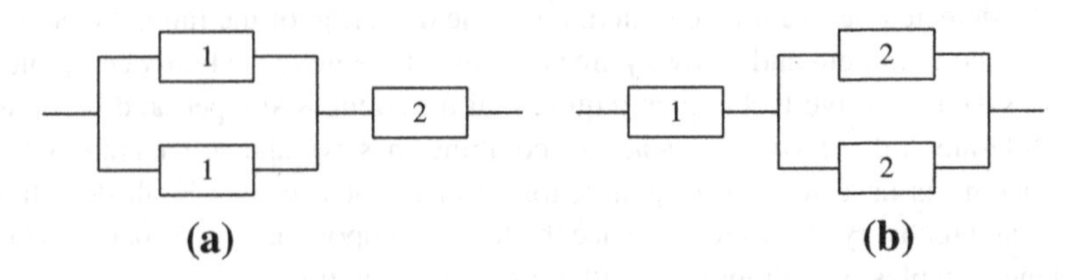

Fig. 4.1 Redundancy allocation

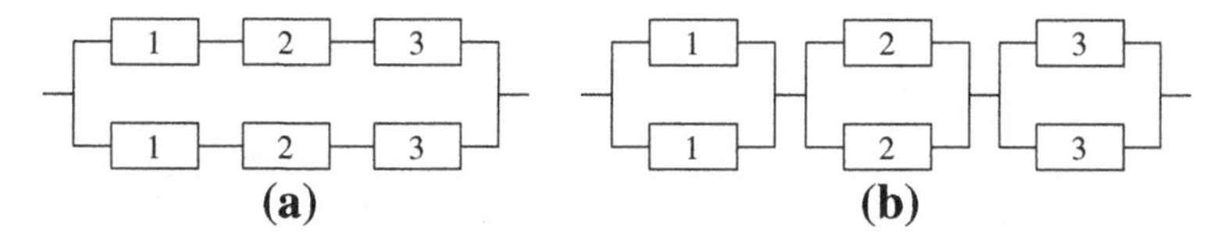

Fig. 4.2 High-level and low-level redundancy

Another important parameter is the *level of redundancy*. Consider a system consisting of three serial components. In high-level redundancy, the entire system in duplicated, as shown in Fig. 4.2a. In low-level redundancy, the duplication takes place at component level, as shown in Fig. 4.2b.

Let us compare the reliabilities of the systems in Fig. 4.2a, b. Assuming that the component failures are mutually independent, we have

$$R_a = 1 - (1 - R_1 R_2 R_3)^2$$
$$R_b = (1 - (1 - R_1)^2)(1 - (1 - R_2)^2)(1 - (1 - R_3)^2).$$

As we can see, the reliabilities R_a and R_b differ, although the systems have the same number of components. If $R_1 = R_2 = R_3 = R$, then the difference is

$$R_b - R_a = 6R^3(1 - R)^2.$$

Consequently, $R_b > R_a$, i.e. low-level redundancy yields a higher reliability than high-level redundancy.

It is important to stress that the conclusion we derived holds only in situations in which component failures are truly independent in both configurations. In reality, in low-level redundancy the redundant components are usually less isolated physically, and therefore are more prone to common sources of stress. Therefore, common-mode failures are more likely to occur at low level rather than at high level [15].

4.2 Passive Redundancy

The passive redundancy approach masks faults rather than detects them. Masking insures that only correct values are passed to the system output in spite of the presence of a fault. Passive redundancy techniques are usually used in high-reliability applications in which even short interruptions of system operation are unacceptable, or in which it is not possible to repair the system. Examples of such applications include aircraft flight control systems, embedded medical devices such as heart pacemakers, and deep-space electronics.

In this section, we first study the concept of triple modular redundancy, and then extend it to a more general case of N-modular redundancy.

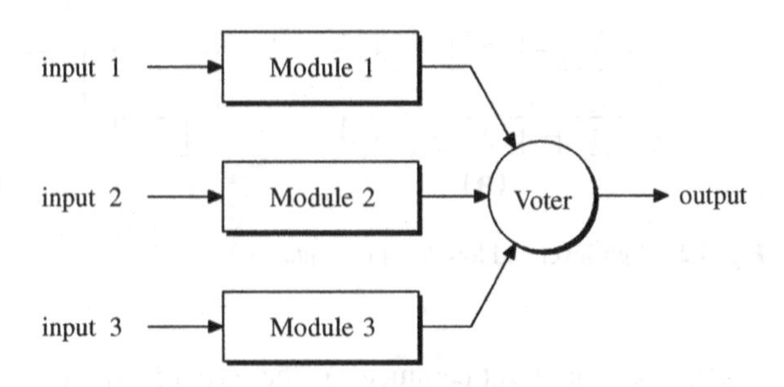

Fig. 4.3 Triple modular redundancy

4.2.1 Triple Modular Redundancy

The most common form of passive hardware redundancy is *Triple Modular Redundancy* (TMR) [16]. The basic configuration is shown in Fig. 4.3. The components are triplicated to perform the same computation in parallel. Majority voting is used to determine the correct result. If one of the modules fails, the majority voter masks the fault by recognizing as correct the result of the remaining two fault-free modules. Depending on the application, the triplicated modules can be processors, memories, disk drives, buses, network connections, power supplies, and so on.

A TMR system can mask only one module fault. A failure in either of the remaining modules would cause the voter to produce an erroneous result. In Sect. 4.4 we show how the dependability of a TMR system can be improved by removing failed modules from the system.

TMR is usually used in applications where a substantial increase in reliability is required for a short period. For example, the use of TMR in the logic section of the launch vehicle digital computer of the Saturn V rocket enabled the reliability of the logic section for a 250-h mission to be increased to 20 times that of an equivalent nonredundant system [4]. Saturn V was used by the National Aeronautics and Space Administration (NASA) agency of the United States Government during 1967–1973 to carry Apollo spacecraft and the Skylab space station into orbit [2].

TMR at a low level is also used to enhance device tolerance to transient faults caused by radiation. For example, some Field Programmable Gate Arrays (FPGAs) intended for space or other radiation-intensive applications employ TMR at the flip-flop level [12].

4.2.1.1 Reliability Evaluation

The fact that a TMR system can mask one module fault does not immediately imply that the reliability of a TMR system is higher than the reliability of a nonredundant system. To estimate the influence of TMR on reliability, we need to take into account the reliability of modules as well as mission time. First, we consider the case when the voter is *perfect*, i.e. it cannot fail.

Perfect Voter

A TMR system operates correctly as long as at least two modules operate correctly. Assuming that the voter is perfect and that the module failures are mutually independent, the reliability of a TMR systems is given by

$$R_{\text{TMR}} = R_1R_2R_3 + (1 - R_1)R_2R_3 + R_1(1 - R_2)R_3 + R_1R_2(1 - R_3),$$

where R_i is the reliability of module M_i, for $i \in \{1, 2, 3\}$. The term $R_1R_2R_3$ is the probability that the first module operates correctly *and* the second module operates correctly *and* the third module operates correctly. The term $(1 - R_1)R_2R_3$ stands for the probability that the first module has failed *and* the second module operates correctly *and* the third module operates correctly, etc. The overall probability is an *or* of the probabilities of the terms since the events are mutually exclusive. If $R_1 = R_2 = R_3 = R$, the above equation reduces to

$$R_{\text{TMR}} = 3R^2 - 2R^3. \quad (4.1)$$

Figure 4.4 compares the reliability of a TMR system, R_{TMR}, to the reliability of a nonredundant system consisting of a single module with reliability R. The reliabilities of the modules composing the TMR system are assumed to be equal to R. As we can see, there is a point at which $R_{\text{TMR}} = R$. This point can be found by solving the equation $3R^2 - 2R^3 = R$. The three solutions are 0.5, 1, and 0 implying that the reliability of a TMR system is equal to the reliability of a nonredundant system when the reliability of the module is $R = 0.5$, when the module is perfect ($R = 1$), or when the module is failed ($R = 0$).

This further illustrates the difference between fault tolerance and reliability. A system can be fault-tolerant and still have a low overall reliability. For example, a TMR system constructed from poor-quality modules with $R = 0.2$ has a low

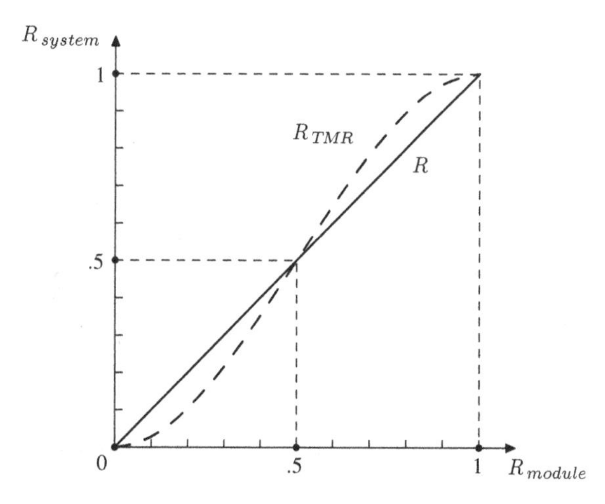

Fig. 4.4 TMR reliability compared to nonredundant system reliability

reliability of $R_{TMR} = 0.136$. Conversely, a system which cannot tolerate any faults may have a high reliability, e.g. when all its components are highly reliable. However, such a system will fail as soon as the first fault occurs.

Next, let us consider how the reliability of a TMR system changes as a function of time. According to exponential failure law (3.10), for a constant failure rate λ, the reliability of the system decreases exponentially with time as $R(t) = e^{-\lambda t}$. Substituting this expression into (4.1), we get

$$R_{TMR}(t) = 3e^{-2\lambda t} - 2e^{-3\lambda t}. \tag{4.2}$$

Figure 4.5 compares the reliabilities of a nonredundant system and a TMR system. The value of normalized time, λt, rather than t is shown on the x-axis, to make the comparison independent of the failure rate. Since, by Eq. (3.14), $1/\lambda = \text{MTTF}$, the point $\lambda t = 1$ corresponds to the time when a system is expected to experience its first failure. We can see that the reliability of a TMR system is higher than the reliability of a nonredundant system in the period between 0 and approximately $\lambda t = 0.69$. Therefore, TMR is suitable for applications whose mission time is shorter than 0.69 of MTTF.

Example 4.1. A controller with an MTTF of 11×10^3 h is expected to operate continuously on a 500-h mission. Check if the mission time can be doubled without decreasing the mission reliability if three controllers are connected in a TMR configuration. Assume that the voter is perfect.

First, we determine the reliability of a single controller for a 500-h mission. The failure rate of a controller is $\lambda = 1/\text{MTTF} = 0.9 \times 10^{-4}$ per hour. Thus, its reliability for a 500-h mission is

$$R(500\,\text{h}) = e^{-\lambda t} = e^{-0.9 \times 10^{-4} \times 500} = 0.956.$$

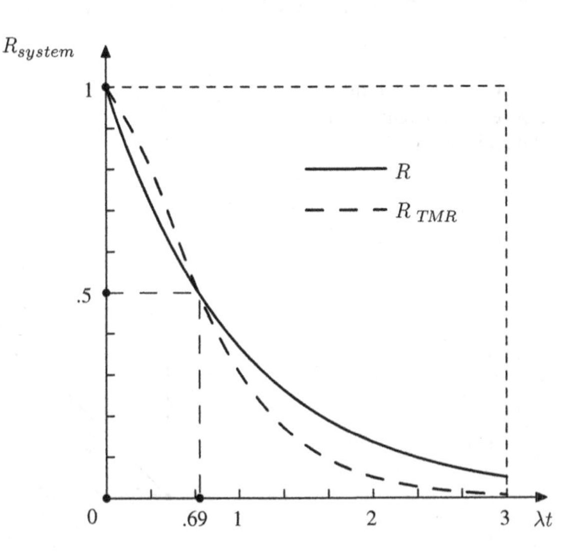

Fig. 4.5 TMR reliability as a function of normalized time λt

Using the Eq. (4.2), we compute the reliability of a TMR configuration for a 1,000-h mission as

$$R_{\mathrm{TMR}}(1{,}000\,\mathrm{h}) = 3e^{-2\lambda t} - 2e^{-3\lambda t} = 0.979.$$

So, the mission time can be doubled without decreasing the mission reliability.

Non-Perfect Voter

In the previous section, we evaluated the reliability of a TMR system assuming that a voter is perfect. Clearly, such an assumption is not realistic. To estimate the reliability of a TMR system more precisely, we need to take the reliability of the voter into account.

The voter is in series with the redundant modules, since if it fails, the whole system fails. Therefore, we get the following expression for reliability of the TMR system with a non-perfect voter:

$$R_{\mathrm{TMR}} = (3R^2 - 2R^3)R_v \tag{4.3}$$

The reliability of the voter must be very high to make the overall reliability of the TMR system higher than the reliability of a nonredundant system. From $(3R^2 - 2R^3)R_v > R$, we can conclude that the reliability of the voter should be at least

$$R_v > \frac{1}{3R - 2R^2}. \tag{4.4}$$

It can be shown [14] that $R = 0.75$ maximizes the value of the expression $3R - 2R^2$. By substituting $R = 0.75$ into Eq. (4.4), we get the minimal value of R_v satisfying Eq. (4.4), which is $R_v = 0.889$.

Fortunately, the voter is typically a very simple device compared to the redundant components, and therefore its failure probability is much smaller. Still, in some systems, the presence of a single point of failure is not acceptable by requirement specifications. The *single point of failure* is any component within a system whose failure leads to the failure of the system. In such cases, schemes with redundant voters are used.

One possibility is to triplicate voters as shown in Fig. 4.6. Such a structure avoids the single point of failure, but requires consensus to be established among three voters. In order to reach consensus, the voters typically exchange several rounds of messages.

Multiple stages of TMR with triplicated voters can be connected as shown in Fig. 4.7. Such a technique is used, for example, in Boeing 777 aircraft to protect all essential hardware resources, including computing systems, electrical power, hydraulic power, and communication paths [17].

Another possibility is to replace a failed voter with a standby voter. In this case, an additional fault-detection unit is required to detect primary voter failure.

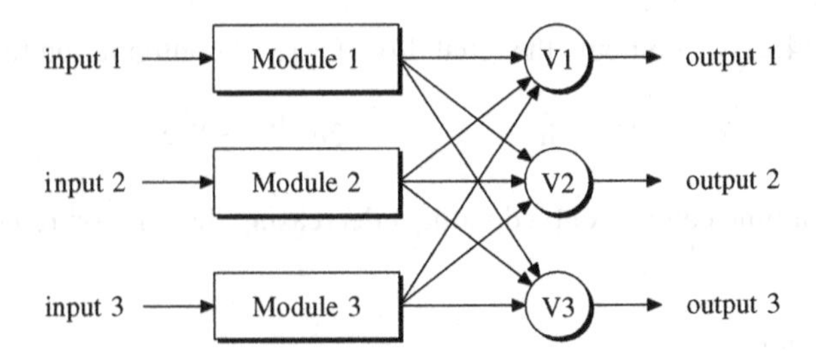

Fig. 4.6 TMR system with three voters

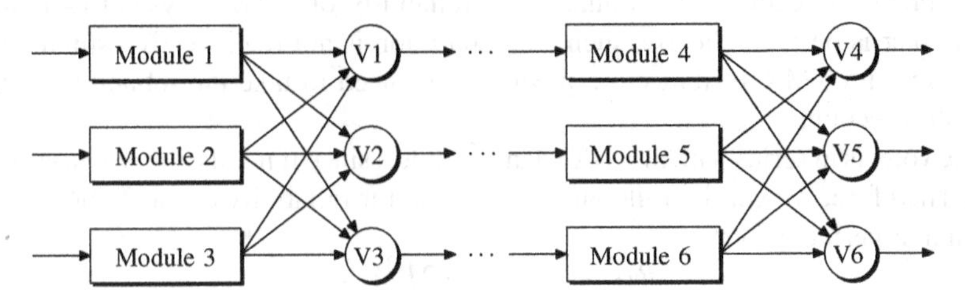

Fig. 4.7 A multiple-stage TMR system

Example 4.2. Repeat Example 4.1 assuming that the MTTF of the voter is 25×10^3 h.

The failure rate of the voter is $\lambda = 1/\text{MTTF} = 0.4 \times 10^{-4}$ per hour. Thus, its reliability for a 1,000-h mission is

$$R(1{,}000\,\text{h}) = e^{-\lambda t} = e^{-0.04} = 0.961.$$

The reliability of a TMR configuration with a non-perfect voter is a product of the reliability of a TMR with a perfect voter and the reliability of the voter. In Example 4.1, we determined that the reliability of a TMR with a perfect voter for a 1,000-h mission is 0.979. So, we get

$$R_{\text{TMR}}(1{,}000\,\text{h}) = 0.979 \times 0.961 = 0.941.$$

So, the mission time cannot be doubled without decreasing the mission reliability.

4.2.1.2 Synchronization and Adjudication

Voting relies heavily on accurate timing [11]. If input values arrive at a voter at different times, an incorrect voting result may be generated. Therefore, a reliable synchronization service should be provided throughout a TMR or NMR systems. This is typically done either by using additional timers, or by implementing asynchronous protocols that rely on the progress of computation to provide an estimate of time.

Multiple-processor systems usually use a fault-tolerant global clock service that maintains a consistent source of time throughout the system.

Another problem with voting is that the values that arrive at a voter may not completely agree, even in a fault-free case. For example, an analog-to-digital converter may produce values which slightly disagree. A common approach to overcoming this problem is to accept as correct the median value which lies between the remaining two. Such a technique is called *median value select*. Another approach is to ignore several least significant bits of information and to perform voting only on the remaining bits. The number of bits that can be safely ignored depends on the application. A comprehensive discussion of various voting techniques can be found in [9].

4.2.1.3 Voter Implementation

Voting can be implemented in either hardware or software. Hardware voters are usually quick enough to meet any response deadline. If voting is done by software voters, they might not be able to reach a consensus within the required time.

A hardware majority voter with three inputs is shown in Fig. 4.8. The value of its output function $f(x_1, x_2, x_3)$ is determined by the majority of values of the input variables x_1, x_2, x_3. The truth table of $f(x_1, x_2, x_3)$ is given in Table 4.1.

4.2.2 N-Modular Redundancy

The N-modular redundancy (NMR) approach is based on the same principle as TMR, but it uses n modules instead of three. The configuration is shown in Fig. 4.9. The

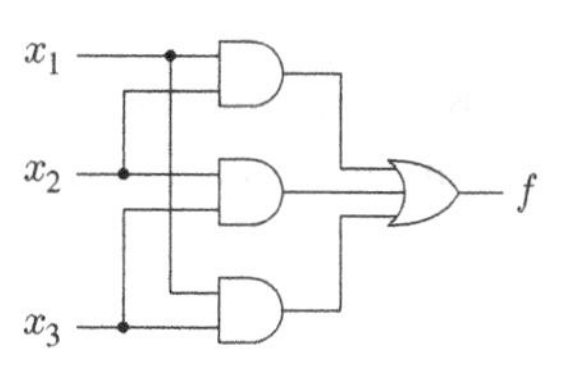

Fig. 4.8 Logic circuit implementing a 2-out-of-3 majority voter

Table 4.1 Truth table of 2-out-of-3 majority voter

x_1	x_2	x_3	f
0	0	0	0
0	0	1	0
0	1	0	0
0	1	1	1
1	0	0	0
1	0	1	1
1	1	0	1
1	1	1	1

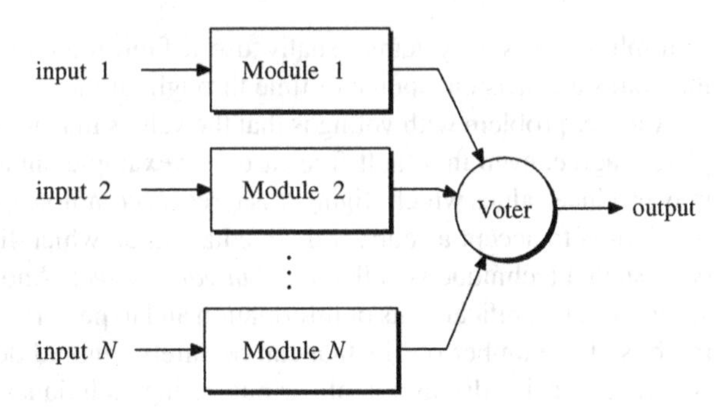

Fig. 4.9 N-modular redundancy

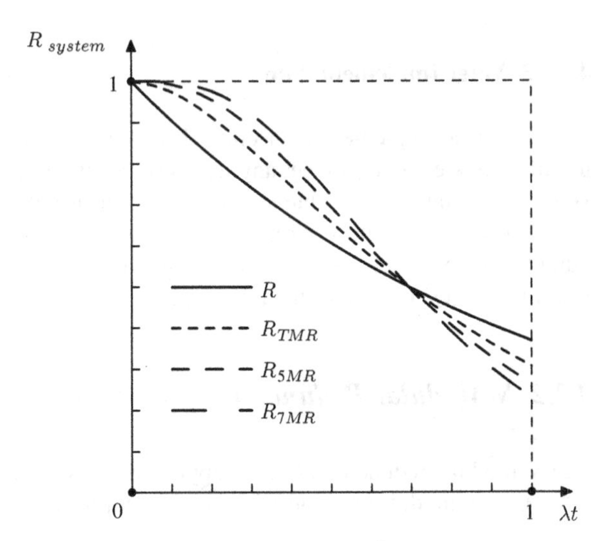

Fig. 4.10 Reliability of an NMR system for different values of N

number N is selected to be odd, to make the majority voting possible. An NMR system can mask $\lfloor N/2 \rfloor$ module faults.

Assuming that the voter is perfect and that the module failures are mutually independent, the reliability of an NMR system with $N = 2n + 1$ is given by

$$R_{\text{NMR}} = \sum_{i=n+1}^{2n+1} \binom{2n+1}{i} R^i (1-R)^{2n+1-i}, \tag{4.5}$$

where R is the reliability of the module. This formula follows directly from the Eq. (3.9).

Figure 4.10 shows the reliabilities of NMR systems for $N = 1, 3, 5$ and 7. As expected, larger values of N result in a higher increase in the reliability of the system. At time approximately $\lambda t = 0.69$, the reliabilities of nonredundant, TMR, 5MR, and 7MR systems become equal. After $\lambda t = 0.69$, the reliability of a nonredundant system is higher than the reliabilities of redundant systems. So, similarly to TMR, NMR is suitable for applications whose mission time is shorter than 0.69 of MTTF.

Example 4.3. Repeat Example 4.2 assuming that 5MR rather than TMR is used.

From the Eq. (4.5), we get the following expression for reliability of 5MR system for the case when the voter is perfect:

$$R_{5MR} = R^5 + 5R^4(1 - R) + 10R^3(1 - R)^2.$$

If the voter is nonperfect, the above expression is multiplied by the reliability of the voter:

$$R_{5MR} = (R^5 + 5R^4(1 - R) + 10R^3(1 - R)^2)R_v.$$

In Example 4.1, we estimated that the reliability of a single controller for 1,000-h mission is $R(1{,}000\,\mathrm{h}) = 0.956$. In Example 4.2, we derived that the reliability of a voter for a 1,000-h mission is $R_v(1{,}000\,\mathrm{h}) = 0.961$. By substituting these values into the above expression, we get the following reliability of a 5MR system for a 1,000-h mission:

$$R_{5MR}(1{,}000\mathrm{h}) = 0.96.$$

So, the mission time can be doubled without decreasing the mission reliability.

4.3 Active Redundancy

Active redundancy achieves fault tolerance by first detecting the faults which occur, and then performing the actions needed to return the system back to an operational state. Active redundancy techniques are typically used in applications requiring high availability, such as time-shared computing systems or transaction processing systems, where temporary erroneous results are preferable to the high degree of redundancy required for fault masking. Infrequent, occasional errors are allowed, as long as the system returns back to normal operation within a specified period of time.

In this section, we consider three common active redundancy techniques: duplication with comparison, standby, and pair-and-a-spare.

4.3.1 Duplication with Comparison

The basic form of active redundancy is *duplication with comparison* shown in Fig. 4.11. Two identical modules operate in parallel. Their results are compared using a comparator. If the results disagree, an error signal is generated. Depending on the application, the duplicated modules can be processors, memories, busses, and so on.

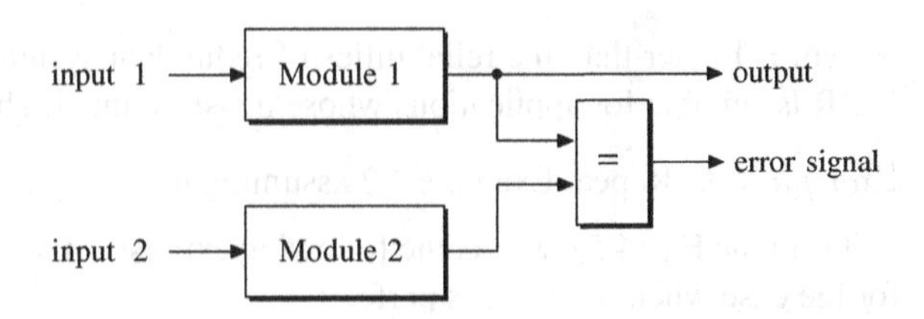

Fig. 4.11 Duplication with comparison

The duplication with a comparison scheme can detect only one module fault. After the fault is detected, no actions are taken by the system to return back to the operational state.

4.3.1.1 Reliability Evaluation

The duplication with a comparison system operates correctly until both modules operate correctly. When the first fault occurs, the comparator detects a disagreement and the normal operation of the system stops. The comparator is not able to distinguish which of the two results is the correct one. Assuming that the comparator is perfect and that the component failures are mutually independent, the reliability of the system is given by

$$R_{DC} = R_1 R_2, \tag{4.6}$$

where R_1 and R_2 are the reliabilities of modules M_1 and M_2, respectively.

Figure 4.12 compares the reliability of a duplication with comparison system with the reliabilities of modules $R_1 = R_2 = R$ to the reliability of a nonredundant system consisting of a single module with the reliability R. We can see that, unless the modules are perfect, i.e. $R(t) = 1$, the reliability of a duplication with comparison system is smaller than the reliability of a nonredundant system.

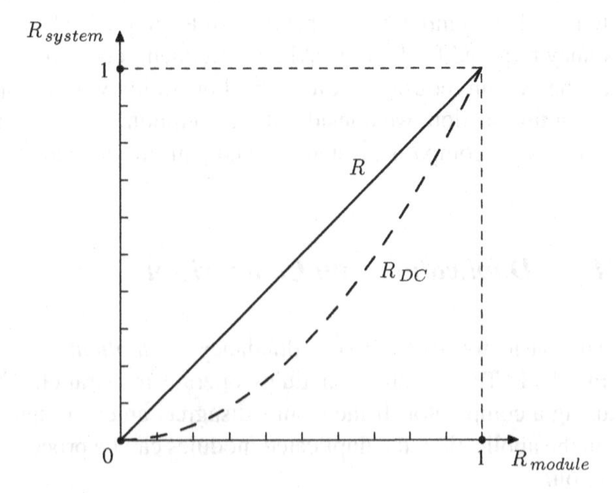

Fig. 4.12 Duplication with comparison reliability compared to nonredundant system reliability

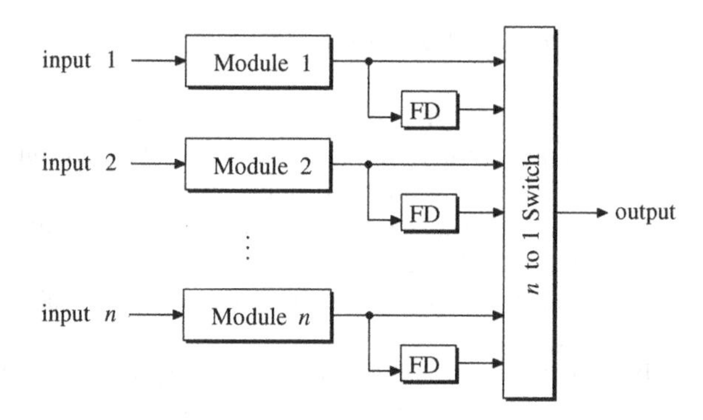

Fig. 4.13 Standby redundancy

4.3.2 Standby Redundancy

Standby redundancy [7] is another technique for active hardware redundancy. The basic configuration is shown in Fig. 4.13. One of the n modules is active. The remaining $n-1$ modules serve as spares (or back-ups). A switch monitors the active module and switches operation to a spare if an error is reported by a fault-detection (FD) unit.

There are two types of standby redundancy: hot standby and cold standby. In the *hot standby*, both operational and spare modules are powered up. Thus, spares can be switched into use immediately after the active module has failed. This minimizes the downtime of the system due to reconfiguration.

In the *cold standby*, the spares are powered down until they are needed to replace a faulty module. This increases reconfiguration time by the amount of time required to power and initialize a spare. However, since spares do not consume power while in the standby mode, such a trade-off is preferable for applications where power consumption is critical, e.g., satellite systems.

A standby redundancy with n modules can tolerate $n-1$ module faults. Here by "tolerate" we mean that a system detects and locates a fault, successfully recovers from it, and continues normal operation. When the nth fault occurs, it is detected by the FD unit. However, since the pool of spares is exhausted, the system is not able to recover and continue its operation. Therefore, the nth fault is not tolerated.

Standby redundancy is used in many applications. One example is the Apollo spacecraft telescope mount pointing computer which was used in Skylab, the first American space station [5]. In this system, two identical computers, an active and a spare, are connected to a switch that monitors the active computer and shifts operation to the spare in the case of a malfunction.

Another example is the NonStop Himalaya server [3]. This system is composed of a cluster of processors working in parallel. Each processor has its own memory and copy of the operating system. A primary process and a spare process are run on separate processors. The spare process mirrors all the information in the primary process and is able to immediately take over in the case of a primary processor failure.

4.3.2.1 Reliability Evaluation

By their nature, standby systems involve dependency between components, since the spare units are held in reserve and only brought into operation if the primary unit fails. Therefore, standby systems are best analyzed using Markov models. We first consider an idealized case where the FD units and the switch cannot fail and the FD units provide the FD coverage 1. Thus, if a fault occurs within a module, the module is successfully replaced by a spare. Later, we also consider the possibility of FD coverage being smaller than 1.

It is important to distinguish between the *failure* of a FD unit and the fault detection *coverage* it provides. The FD coverage is determined by the FD mechanism implemented in the FD unit. For example, a certain *known* percentage of faults in a module is undetectable by the error-detecting code which is used in the FD unit. So, if the FD coverage is smaller than 1, it does not mean that the FD unit does not operate as expected. On the contrary, if a FD unit *fails*, it means that it has ceased to perform its functions correctly. Thus, it delivers erroneous results which, at some point, may cause a failure of the system.

Fault Detection Coverage 1

Consider a standby redundancy scheme with one spare, shown in Fig. 4.14. Let module 1 be a primary module and module 2 be a spare. Suppose also that we have a case of cold standby, i.e. the spare is powered down until it is needed to replace a faulty module. In this case, it is natural to assume that the spare cannot fail while in standby mode. This assumption is usually used in the analysis of cold standby systems.

A Markov chain of the system for the reliability evaluation with no repairs is shown in Fig. 4.15. Its states are numbered according to Table 4.2.

When the primary component fails, there is a transition between state 1 and state 2. If the system is in state 2 and the spare fails, there is a transition to state 3. Since we assumed that the spare cannot fail while in standby mode, the combination (O, F) cannot occur. The states 1 and 2 are operational states. The state 3 is the failed state.

The transition matrix for the Markov chain in Fig. 4.15 is given by

$$\mathbf{M} = \begin{bmatrix} -\lambda_1 & 0 & 0 \\ \lambda_1 & -\lambda_2 & 0 \\ 0 & \lambda_2 & 0 \end{bmatrix}.$$

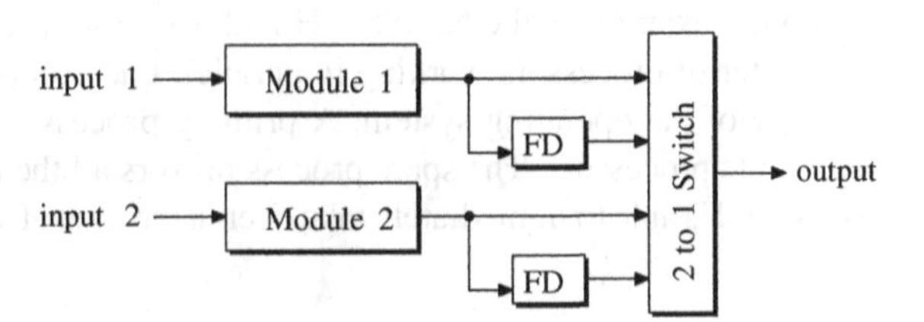

Fig. 4.14 Standby redundancy with one spare

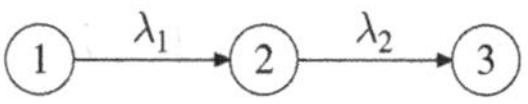

$$\textcircled{1} \xrightarrow{\lambda_1} \textcircled{2} \xrightarrow{\lambda_2} \textcircled{3}$$

Fig. 4.15 Markov chain of a cold standby redundancy with one spare

Table 4.2 States of the Markov chain in Fig. 4.15

Component		State number
1	2	
O	*O*	1
F	*O*	2
F	*F*	3

So, we get the following system of state transition equations:

$$\frac{\mathrm{d}}{\mathrm{d}t}\begin{bmatrix} P_1(t) \\ P_2(t) \\ P_3(t) \end{bmatrix} = \begin{bmatrix} -\lambda_1 & 0 & 0 \\ \lambda_1 & -\lambda_2 & 0 \\ 0 & \lambda_2 & 0 \end{bmatrix} \times \begin{bmatrix} P_1(t) \\ P_2(t) \\ P_3(t) \end{bmatrix}$$

or

$$\begin{cases} \frac{\mathrm{d}}{\mathrm{d}t} P_1(t) = -\lambda_1 P_1(t) \\ \frac{\mathrm{d}}{\mathrm{d}t} P_2(t) = \lambda_1 P_1(t) - \lambda_2 P_2(t) \\ \frac{\mathrm{d}}{\mathrm{d}t} P_3(t) = \lambda_2 P_2(t). \end{cases}$$

By solving this system of equations, we get

$$\begin{aligned} P_1(t) &= e^{-\lambda_1 t} \\ P_2(t) &= \frac{\lambda_1}{\lambda_2 - \lambda_1}(e^{-\lambda_1 t} - e^{-\lambda_2 t}) \\ P_3(t) &= 1 - \frac{1}{\lambda_2 - \lambda_1}(\lambda_2 e^{-\lambda_1 t} - \lambda_1 e^{-\lambda_2 t}). \end{aligned}$$

The reliability of the system is the sum of $P_1(t)$ and $P_2(t)$:

$$R_{SS}(t) = e^{-\lambda_1 t} + \frac{\lambda_1}{\lambda_2 - \lambda_1}(e^{-\lambda_1 t} - e^{-\lambda_2 t}).$$

This can be rewritten as

$$R_{SS}(t) = e^{-\lambda_1 t} + \frac{\lambda_1}{\lambda_2 - \lambda_1} e^{-\lambda_1 t}(1 - e^{-(\lambda_2 - \lambda_1)t}). \tag{4.7}$$

Assuming $(\lambda_2 - \lambda_1)t \ll 1$, we can expand the term $e^{-(\lambda_2 - \lambda_1)t}$ as a power series of $-(\lambda_2 - \lambda_1)t$ as

$$e^{-(\lambda_2-\lambda_1)t} = 1 - (\lambda_2 - \lambda_1)t + 1/2(\lambda_2 - \lambda_1)^2 t^2 - \ldots$$

Substituting it into (4.7), we get

$$R_{SS}(t) = e^{-\lambda_1 t} + \lambda_1 e^{-\lambda_1 t}(t - 1/2(\lambda_1 - \lambda_2)t^2 + \ldots).$$

Assuming $\lambda_2 = \lambda_1$, we can simplify the above expression to

$$R_{SS}(t) = (1 + \lambda t)e^{-\lambda t}. \tag{4.8}$$

Next, let us see how the Eq. (4.8) will change if we ignore the dependency between the failures of components. If the primary and spare module failures are treated as mutually independent, the reliability of a standby system is the sum of two probabilities:

1. The probability that module 1 operates correctly, and
2. The probability that module 2 operates correctly, while module 1 has failed and has been replaced by module 2.

Then, we get the following expression for the reliability of the system:

$$R_{SS} = R_1 + (1 - R_1)R_2.$$

If $R_1 = R_2 = R$, then

$$R_{SS} = 2R - R^2$$

or

$$R_{SS}(t) = 2e^{-\lambda t} - e^{-2\lambda t}. \tag{4.9}$$

Figure 4.16 compares the plots of reliabilities given by Eqs. (4.8) and (4.9). We can see that neglecting the dependency between failures leads to underestimating the standby system reliability. In the former case, $R_{SS}(t) = 0.74$, while in the latter case $R_{SS}(t) = 0.60$.

Example 4.4. Draw a Markov chain for availability evaluation of the hot standby redundancy with three modules. Assume that

- the failure rate of each module is λ and the repair rate is μ,
- the failure rate of the switch is λ_s and the repair rate is μ_s,
- the FD unit is perfect and the FD coverage is 1.

The resulting Markov chain is shown in Fig. 4.17. The states are labeled according to Table 4.3. The states 1–3 are operational states. The states 4–8 are failed states. Since we perform availability evaluation, the system can be repaired from a failed state as well.

In a hot standby configuration all modules are powered up. Therefore, spare modules can fail in a standby mode. For this reason, the transition e.g. from states 1 to 2 occurs at the rate 3λ. If we were using cold standby, the rate would be λ.

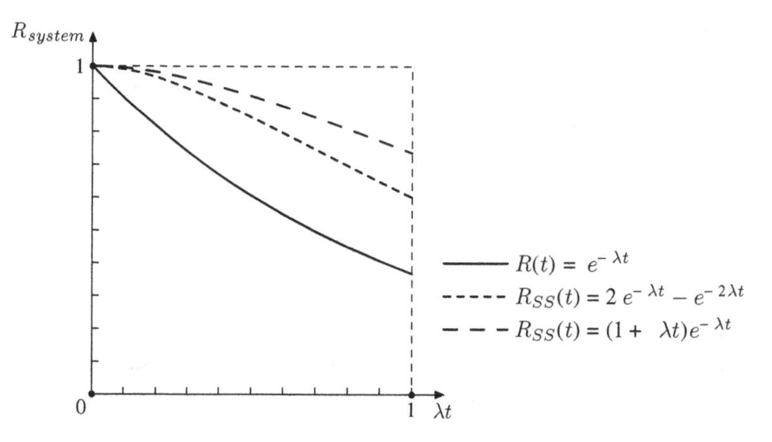

Fig. 4.16 Cold standby redundancy reliability compared to nonredundant system reliability

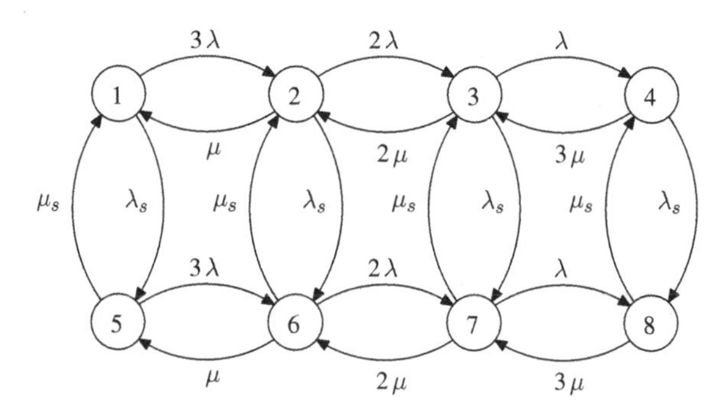

Fig. 4.17 Markov chain for Example 4.4

Table 4.3 States of the Markov chain for Example 4.4

State	Description
1	Three components and switch are operational
2	One component is failed, two components and switch are operational
3	Two components are failed, one component and switch are operational
4	Three components are failed, switch is operational
5	Three components are operational, switch is failed
6	Two components are operational, one component and switch are failed
7	One component is operational, two components and switch are failed
8	Three components and switch are failed

Fault Detection Coverage Smaller than 1

Next, we consider the case when the FD coverage is smaller than 1. Suppose that the probability that the fault is detected is C. Then, the probability that the fault is

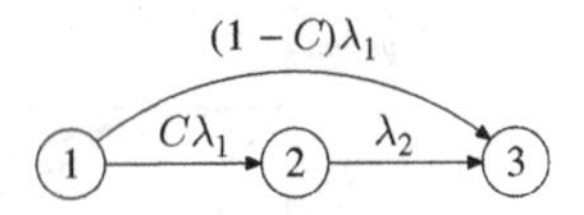

Fig. 4.18 State transition diagram of a cold standby redundancy with one spare and fault detection coverage C

not detected is $1 - C$. We assume that if a fault is detected, the switch successfully replaces the primary unit by a spare. If a fault is not detected, the system fails.

The Markov chain with these assumptions for the reliability evaluation is shown in Fig. 4.18. The transition from state 1 is partitioned into two transitions. The failure rate is multiplied by C to get the rate of successful transition to state 2. The failure rate is multiplied by $1 - C$ to get the rate of failure due to an undetected fault. We assume that no repairs are allowed.

The state transition equations corresponding to the Markov chain in Fig. (4.18) are

$$\begin{cases} \frac{\mathrm{d}}{\mathrm{d}t}P_1(t) = -\lambda_1 P_1(t) \\ \frac{\mathrm{d}}{\mathrm{d}t}P_2(t) = C\lambda_1 P_1(t) - \lambda_2 P_2(t) \\ \frac{\mathrm{d}}{\mathrm{d}t}P_3(t) = \lambda_2 P_2(t) + (1-C)\lambda_1 P_1(t). \end{cases}$$

By solving this system of equations, we get

$$P_1(t) = e^{-\lambda_1 t}$$

$$P_2(t) = \frac{C\lambda_1}{\lambda_2 - \lambda_1}(e^{-\lambda_1 t} - e^{-\lambda_2 t})$$

$$P_3(t) = 1 - \left(1 + \frac{C\lambda_1}{\lambda_2 - \lambda_1}\right)e^{-\lambda_1 t} + \frac{C\lambda_1}{\lambda_2 - \lambda_1}e^{-\lambda_2 t}.$$

As before, state 3 is a failed state. So, the reliability of the system is the sum of $P_1(t)$ and $P_2(t)$:

$$R_{SS}(t) = e^{-\lambda_1 t} + \frac{C\lambda_1}{\lambda_2 - \lambda_1}(e^{-\lambda_1 t} - e^{-\lambda_2 t}).$$

Assuming $\lambda_2 = \lambda_1$, the above can be simplified to:

$$R_{SS}(t) = (1 + C\lambda t)e^{-\lambda t}. \tag{4.10}$$

Figure 4.19 compares the reliability of a cold standby redundancy for different values of FD coverage C. As C decreases, the reliability of a standby redundancy decreases. When C reaches zero, a standby redundancy reliability reduces to the reliability of a nonredundant system.

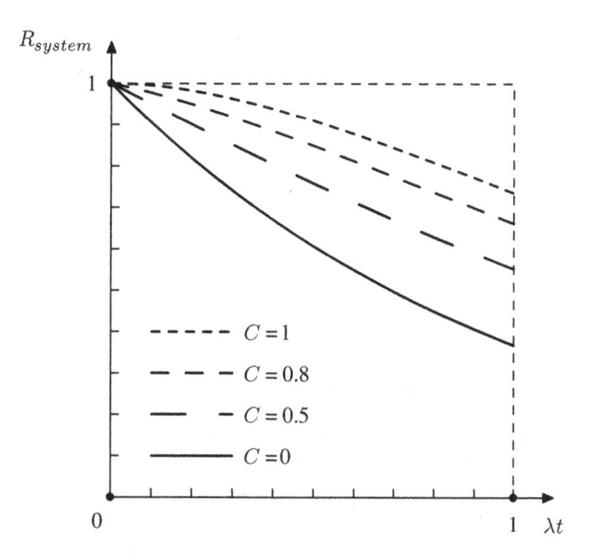

Fig. 4.19 Reliability of a cold standby redundancy for different values of fault detection coverage C

Example 4.5. Two identical components are connected in a cold standby configuration. Compute the reliability of this system during a 3,000-h period if the failure rate of a single component is $\lambda = 0.4 \times 10^{-4}$ per hour and the FD coverage is 97 %. Assume that

- FD units and switch cannot fail,
- the spare component cannot fail while in the standby mode,
- no repairs are allowed.

Using the Eq. 4.10, we get:

$$R_{SS}(3,000\,\text{h}) = (1 + C\lambda t)e^{-\lambda t} = (1 + 0.97 \times 0.04 \times 3)e^{-0.04 \times 3} = 0.99.$$

Example 4.6. Draw a Markov chain for safety evaluation of the cold standby redundancy with three modules. Assume that the failure rate of each module is λ and the repair rate is μ, the spares cannot fail while in the standby mode, the switch and FD units are perfect, and the FD coverage is C. Also assume that the system cannot be repaired if it failed nonsafe; otherwise, it can be repaired.

The resulting Markov chain is shown in Fig. 4.20. The states are labeled according to Table 4.4. States 1–3 are operational states. State 4 is a failed-safe state. State 5 is a failed-unsafe state.

4.3.3 Pair-And-A-Spare

The pair-and-a-spare technique [6] combines the standby redundancy and the duplication and comparison approaches. Its configuration is shown in Fig. 4.21.

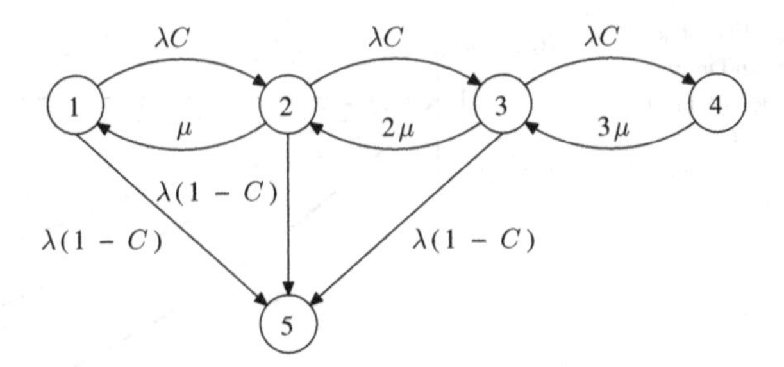

Fig. 4.20 Markov chain for Example 4.6

Table 4.4 States of the Markov chain for Example 4.6

State	Description
1	Three components are operational
2	One component is failed and its failure is detected, two components are operational
3	Two components are failed and their failures are detected, one component is operational
4	Three components are failed and their failures are detected (fail-safe state)
5	The failure of a component is not detected (fail-unsafe state)

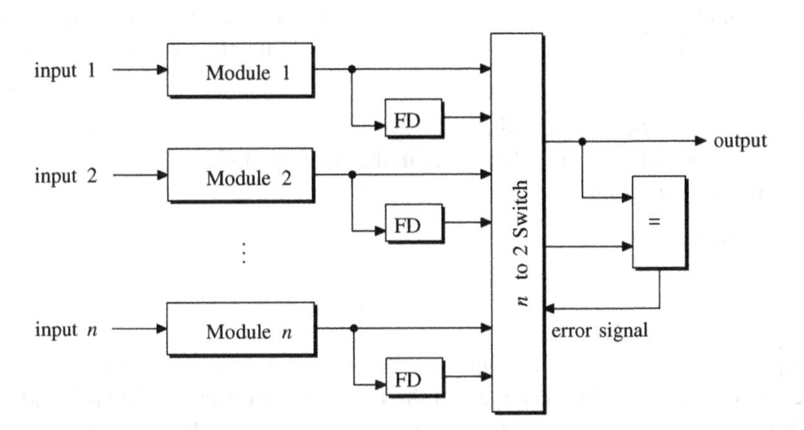

Fig. 4.21 Pair-and-a-spare redundancy

The idea is similar to standby redundancy except that two active modules instead of one are operated in parallel. As in the duplication with comparison case, the results are compared to detect disagreement. If an error signal is received from the comparator, the switch analyzes the report from the FD blocks and decides which of the two modules is faulty. The faulty module is removed from operation and replaced with a spare module.

A pair-and-a-spare system with n modules can tolerate $n-2$ module faults. When the $n-1$st fault occurs, it will be detected and located by the switch and the correct result will be passed to the system's output. However, since there will be no more spares available, the switch will not be able to replace the faulty module with a spare module. Therefore, the system will not be able to continue its normal operation.

Clearly, it is possible to design an extended version of the pair-and-a-spare system in which the system is reconfigured to operate with one active module after $n-1$st fault. Upon receiving a signal that there are no spares left, the comparator can be disconnected and the switch can be modified to an n-to-1 switch which receives an error signal from the fault detection unit, as in the standby redundancy case. Such an extended pair-and-a-spare system would tolerate $n-1$ module faults and detect n module faults.

Example 4.7. Draw a Markov chain for reliability evaluation of the pair-and-a-spare redundancy with four modules. Assume that

- the failure rate of each module is λ and the repair rate is μ,
- the failure rate of the comparator is λ_c,
- the FD unit and the switch are perfect,
- the fault detection coverage is 1.

The resulting Markov chain is shown in Fig. 4.22. The states are labeled according to Table 4.5. States 1–3 are operational states. State 4 is a failed state. In a pair-and-a-spare configuration, we need at least two modules for the system to be operational. Thus, when the third component fails, the system fails. Since we perform reliability evaluation, the system cannot be repaired from a failed state. Therefore, we do not need to distinguish between failures of a system due to the failure of the comparator or to modules. All failed states can be merged into one.

4.4 Hybrid Redundancy

Hybrid redundancy combines advantages of passive and active approaches. Fault masking is used to prevent the system from producing momentary erroneous results. FD, location, and recovery are used to reconfigure a system when a fault occurs. Hybrid redundancy techniques are usually used in safety-critical applications such as control systems for chemical processes, nuclear power plants, weapons, medical equipment, aircrafts, trains, automobiles, and so on.

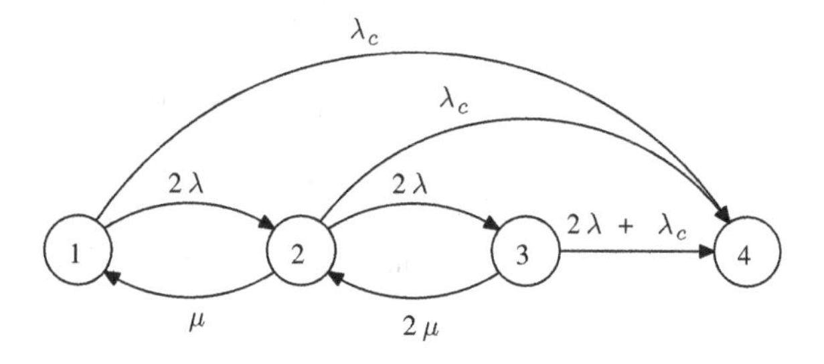

Fig. 4.22 Markov chain for Example 4.7

Table 4.5 States of the Markov chain for Example 4.7

State	Description
1	Four components and the comparator are operational
2	One component is failed, three components and the comparator are operational
3	Two components are failed, two components and the comparator are operational
4	The system failed

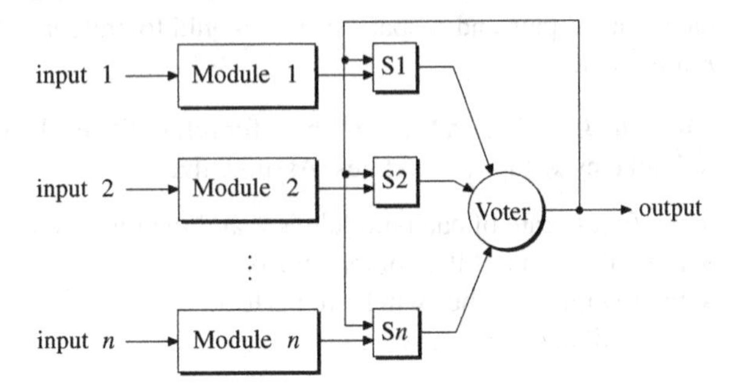

Fig. 4.23 Self-purging redundancy

In this section, we consider two techniques for hybrid redundancy: self-purging redundancy and N-modular redundancy with spares.

4.4.1 Self-Purging Redundancy

Self-purging redundancy [8] consists of n identical modules which perform the same computation in parallel and actively participate in voting (Fig. 4.23). The output of the voter is compared to the outputs of each individual module to detect disagreement. If a disagreement occurs, the switch opens and removes, or *purges*, the faulty module from a system. The voter is designed as a *threshold gate*, capable of adapting to the decreasing number of inputs.

A threshold gate operates as follows [1]. Each of its n inputs $x_1, x_2, \ldots, x_n$ has a weight $w_1, w_2, \ldots, w_n$. The output f is determined by comparing the sum of weighted input values to some threshold value, T:

$$f = \begin{cases} 1, \text{ if } \sum_{i=1}^{n} w_i x_i \geq T \\ 0, \text{ otherwise} \end{cases}$$

where the addition and multiplication are regular arithmetic operations.

The threshold voter removes a faulty module from voting by forcing its weight to zero. In this way, faulty modules do not contribute to the weighted sum.

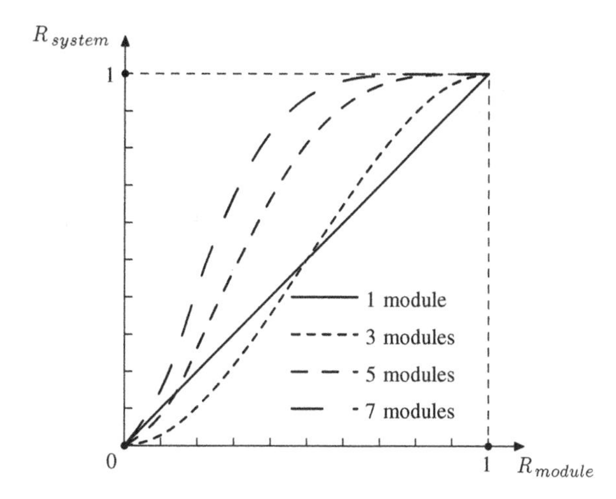

Fig. 4.24 Reliability of a self-purging redundancy system with 3, 5 and 7 modules

A self-purging redundancy system with n modules can mask $n - 2$ module faults. When $n - 2$ modules are purged and only two are left, the system will be able to detect the next, $n - 1$th fault, but, as in the duplication with comparison case, the voter will not be able to distinguish which of the two results is correct.

4.4.1.1 Reliability Evaluation

Since all the modules of the system operate in parallel, we can assume that the failures of the modules are mutually independent. It is sufficient that two of the modules of the system function correctly for the system to be operational. If the voter and the switches are perfect, and all modules have the same reliability $R_1 = R_2 = \ldots = R_n = R$, then the system is *not* reliable if all the modules have failed (probability $(1-R)^n$), or if all modules but one have failed (probability $R(1-R)^{n-1}$). Since there are n choices for one out of n modules to remain operational, we get the equation

$$R_{\mathrm{SP}} = 1 - ((1-R)^n + nR(1-R)^{n-1}). \tag{4.11}$$

Figure 4.24 compares the reliabilities of self-purging redundancy systems with 3, 5, and 7 modules.

It is also interesting to see how the reliability of a self-purging redundancy system compares to the reliability of a passive N modular redundancy with the same number of modules. Figure 4.25 shows the corresponding plots when the number of modules is five. We can see that the reliability of self-purging redundancy with five modules is significantly higher that the reliability of 5MR. For example, for $R = 0.7$, we have $R_{\mathrm{5MR}} = 78\,\%$ and $R_{\mathrm{SP(5)}} = 96.9\,\%$.

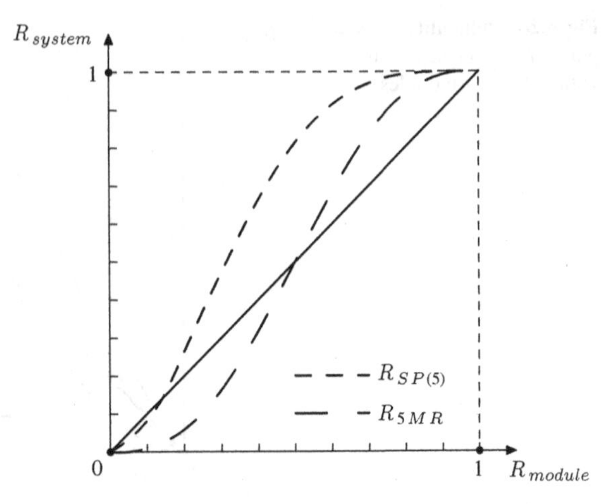

Fig. 4.25 Reliability of a self-purging redundancy system with five modules compared to the reliability of a 5MR system

Example 4.8. A switch has the design life reliability $R = 0.85$. A designer is given the task to increase the design life reliability to at least 0.93. He should choose between a TMR and a self-purging redundancy configuration. Among available components, he has a majority voter with the reliability $R_{\text{vm}} = 0.99$ and a threshold voter with the reliability $R_{\text{vt}} = 0.94$ for the required system design life. Determine which configuration should be used.

Using the Eq. 4.3, we can compute the reliability of the TMR system as

$$R_{\text{TMR}} = (3R^2 - 2R^3)R_{\text{vm}} = 0.939 \times 0.99 = 0.929.$$

From the Eq. 4.11, we can derive an expression for the reliability of self-purging redundancy with three modules. By multiplying it by the reliability of the threshold voter, we get

$$R_{\text{SP(3)}} = (1 - ((1 - R)^3 + 3R(1 - R)^2))R_{vt} = 0.997 \times 0.94 = 0.937.$$

So, to achieve a design life reliability of at least 0.93, self-purging redundancy should be used.

Example 4.9. Determine how large the design life reliability of the threshold voter should be in Example 4.8 if a self-purging redundancy with two modules is to be used.

A self-purging redundancy configuration with two modules is operational if both modules are operational or one of the modules is failed. So, for the case of a perfect voter, we get the following design life reliability:

$$R_{\text{SP(2)}} = R^2 + 2R(1 - R) = 0.977.$$

To achieve $0.977 \times R_{\mathrm{vt}} \geq 0.93$, the design life reliability of the threshold voter should be $R_{\mathrm{vt}} \geq 0.93/0.977 = 0.952$.

4.4.2 N-Modular Redundancy with Spares

In an N-modular redundancy with k spares [13], n modules operate in parallel to provide input to a majority voter (Fig. 4.26). Other k modules serve as spares. If one of the primary modules becomes faulty, the switch replaces the faulty module with a spare one. Various techniques are used to identify faulty modules. One approach is to compare the output of the voter with the individual outputs of the modules, as shown in Fig. 4.26. A module which disagrees with the majority is declared faulty.

After the pool of spares is exhausted, the disagreement detector is switched off and the system continues working as a passive NMR system. Thus, N-modular redundancy with k spares can mask $\lfloor n/2 \rfloor + k$ module faults.

Example 4.10. Draw a Markov chain for reliability evaluation of the three-modular redundancy with two spares. Assume that:

- the failure rate of each of the modules and each of the spares is λ,
- the spares cannot fail while in standby mode,
- the failure rate of the voter is λ_v,
- the switch and disagreement detector are perfect,

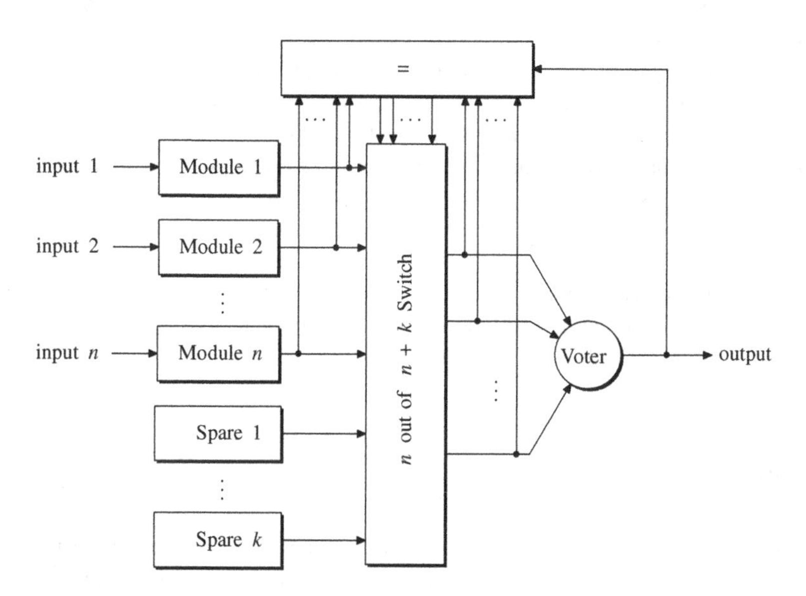

Fig. 4.26 N-modular redundancy with k spares

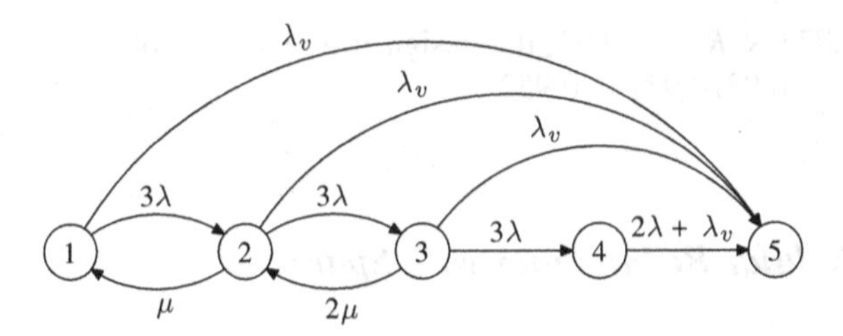

Fig. 4.27 Markov chain for Example 4.10

Table 4.6 States of the Markov chain for Example 4.10

State	Description
1	Three components and two spares are operational
2	One component is failed and replaced by a spare, one spare remains
3	Two components are failed and replaced by spares, no spares remain
4	One component failed in TMR configuration
5	System failed

- only the modules which are identified as faulty by the disagreement detector can be repaired. Once the system is reduced to a passive TMR system and the disagreement detector is switched off, a faulty module cannot be repaired.

The resulting Markov chain is shown in Fig. 4.27. The states are labeled according to Table 4.6. States 1–4 are operational states. State 5 is a failed state. We need at least two modules for the system to be operational. Thus, when three out of five components have failed, the system fails. Since we perform a reliability evaluation, the system cannot be repaired from a failed state. Therefore, we do not need to distinguish between failures of a system due to the failure of the voter or modules. All failed states can be merged into one. Furthermore, due to the last assumption, we cannot repair from state 4. This is a reasonable assumption, because once the system is reduced to a passive TMR system and the disagreement detector is switched off, a faulty module is masked rather than detected. Thus, the repair team does not know if there are any faulty modules to repair. In order to find a faulty module, the TMR system has to be stopped and tested. Since reliability analysis requires continuous operation, system interrupts are not permitted.

4.5 Summary

In this chapter, we have studied different techniques for designing fault-tolerant hardware systems. We have considered three types of hardware redundancy: passive, active, and hybrid. We have explored the most popular redundancy configurations and evaluated their effect on system dependability. We have learned that the increase in complexity redundancy can be quite severe and may diminish the dependability improvement, unless redundant resources are allocated in a proper way.

Problems

In all problems, we assume that failures and repairs of components are independent events (unless it is specified otherwise) and failure and repair rates are constant.

4.1. Suppose that the components in the system shown in Fig. 4.1 have the same cost. Their reliabilities are $R_1 = 0.75$ and $R_2 = 0.96$. Your task is to increase the reliability of the system by adding to it two redundant components. Determine which of the following choices is better:

1. to replace component 1 by a three-component parallel system
2. to replace each of the components 1 and 2 by a two-component parallel system.

4.2. After a year of service, the reliability of a controller is 0.85.

1. If three controllers are connected in a TMR configuration, what is the reliability of the resulting system during the first year of operation? Assume that the voter is perfect and no repairs are allowed.
2. Suppose that the reliability of the voter is $R_v(t)$. Estimate for which value of R_v(1 year) the reliability of the TMR configuration becomes equal to the reliability of a single controller.

4.3. Draw a Markov chain for reliability evaluation of the TMR with three voters shown in Fig. 4.6. Assume that the failure rate of each module is λ_m, and the failure rate of each voter is λ_v. No repairs are allowed.

4.4. Draw a logic diagram of a majority voter with five inputs.

4.5. Suppose a processor has a constant failure rate of $\lambda = 0.184 \times 10^{-3}$ per hour. Which value of N should be used if a system consisting of N such processors in an NMR configuration is to run continuously for 800 h with a failure probability of no more than 4 %? Assume that the voter is perfect and no repairs are allowed.

4.6. A system in the TMR configuration is composed of three modules with reliabilities $R_1 = 0.85$, $R_2 = 0.87$ and $R_2 = 0.89$ during the first year of operation. Compute the reliability of the system during the first year of operation. Assume that the voter is perfect and no repairs are allowed.

4.7. An engineer designs a system consisting of two subsystems in series with the reliabilities $R_1 = 0.99$ and $R_2 = 0.85$. The cost of the two subsystems is the same. The engineer decides to add two redundant components. Which of the following is the best choice:

1. Duplicate subsystems 1 and 2 using high-level redundancy (as in Fig. 4.2a).

2. Duplicate subsystems 1 and 2 using low-level redundancy (as in Fig. 4.2b).
3. Replace the second subsystem by a three-component parallel system.

4.8. Compare the MTTF of a 5MR with the failure rate of each module 0.001 failures per hour with the MTTF of a TMR with the failure rate of each module 0.01 failures per hour. Assume that voters are perfect and no repairs are allowed.

4.9. Draw a Markov chain for reliability evaluation for a 5MR with a failure rate of each module λ. Assume that the voter is perfect and no repairs are allowed.

4.10. A disk drive has a constant failure rate and an MTTF of 7,000 h.

1. What is the probability of failure during the first year of operation?
2. What is the probability of failure during the first year of operation if two of the drives are placed in a cold standby configuration? Assume that:
   - FD units and switch cannot fail,
   - the FD coverage is 1,
   - the spare computer cannot fail while in the standby mode,
   - no repairs are allowed.

4.11. A system has MTTF $= 12 \times 10^3$ h. An engineer is to set the design life time, so that the end-of-life reliability is 0.95.

1. Determine the design life time.
2. Check if the design life time can be doubled without decreasing the end-of-life reliability if two systems are placed in a cold standby redundancy configuration. Make the same assumptions as in Problem 4.10.
3. Check if the design life time can be increased three times without decreasing the end-of-life reliability if three systems are placed in a TMR configuration. Assume that the voter is perfect and no repairs are allowed.

4.12. A computer with an MTTF of 3,000 h is expected to operate continuously on a 1,000-h mission.

1. Compute the computer's reliability during the mission.
2. Suppose two such computers are connected in a cold standby redundancy configuration. Compute the mission reliability of such a system. Make the same assumptions as in Problem 4.10.
3. Determine how the mission reliability will change if the FD coverage is 95 %?

4.13. A chemical process controller has a design life reliability of 0.97. Because the reliability is considered too low, it is decided to duplicate the controller. The design engineer should choose between a parallel and a cold standby redundancy configuration. How large should the fault detection coverage be for the standby configuration to be more reliable than the parallel configuration? For the standby redundancy configuration, assume that FD units and switch cannot fail and the spare controller cannot fail while in the standby mode. No repairs are allowed for either configuration.

4.14. Give examples of applications where you would recommend using cold standby and hot standby redundancy. Justify your answer.

4.15. Draw a Markov chain for safety evaluation for a hold standby redundancy with four modules. Assume that, at each step, the switch examines the reports from all four FD units. If an error is reported, the faulty module is switched off. Also assume that:

- the failure rate of each module is λ and the repair rate is μ,
- the FD unit and switch cannot fail,
- the FD coverage is C,
- the system cannot be repaired if it failed nonsafe; otherwise, it can be repaired,
- if the system failed nonsafe, it cannot fail further (for example, because it ceases to exist).

4.16. Draw a Markov chain for reliability evaluation for a self-purging redundancy with three modules. Assume that:

- the failure rate of each module is λ,
- the failure rate of the voter is λ_v,
- the switches are perfect,
- no repairs are allowed.

4.17. Repeat Problem 4.16 for reliability evaluation assuming that:

- the failure rate of each module is λ and the repair rate is μ,
- the voter and the switches are perfect.

4.18. Repeat Problem 4.16 for availability evaluation. Use the same assumptions as in Problem 4.17.

4.19. Draw a Markov chain for availability evaluation for a self-purging redundancy with five modules. Assume that:

- the failure rate of each module is λ and the repair rate is μ,
- the failure rate of the voter is λ_v and the repair rate is μ_v,
- switches are perfect.

4.20. Draw a Markov chain for reliability evaluation of a five-modular redundancy with two spares. Assume that:

- the failure rate of each module and spare is λ,
- disagreement detector, switch and voter are perfect,
- the spare modules cannot fail while in the standby mode.
- no repairs are allowed.

4.21. Suppose that we replace the majority voter in N-modular redundancy with k spares (Fig. 4.26) by a threshold voter similar to the one used in self-purging redundancy. In this case, the disagreement detector is not switched off after the pool of spares is exhausted. The system continues working as a passive NMR system and the disagreement detector continues identifying faulty modules by comparing the output of the voter with the individual outputs of the modules.

When it identifies a faulty module, it is removed from voting by forcing its weight to 0.
How many module faults can be tolerated in such a system?

4.22. Draw a Markov chain for reliability evaluation of the three-modular redundancy with one spare. Assume that:

- the failure rate of each of the three main modules is λ_m,
- the failure rate of the spare is λ_s,
- the spare cannot fail while in the standby mode,
- the voter, switch and disagreement detector are perfect,
- no repairs are allowed.

4.23. Repeat Problem 4.22 for availability evaluation with the following assumptions:

- the modules and the spare have the same failure rate λ and the same repair rate μ,
- the voter has the failure rate λ_v and the repair rate is μ_v,
- the spare cannot fail while in the spare mode.
- the switch and disagreement detector units are perfect.
- if the system fails, it shuts itself down and cannot fail further.

4.24. The configuration shown in Fig. 4.28 is called triplex–duplex redundancy [6]. A total of six identical modules, grouped in three pairs, are operating in parallel. In each pair, the results of the computation are compared using a comparator. If the results agree, the output of the comparator participates in the voting. Otherwise, the pair of modules is declared faulty and the switch removes the

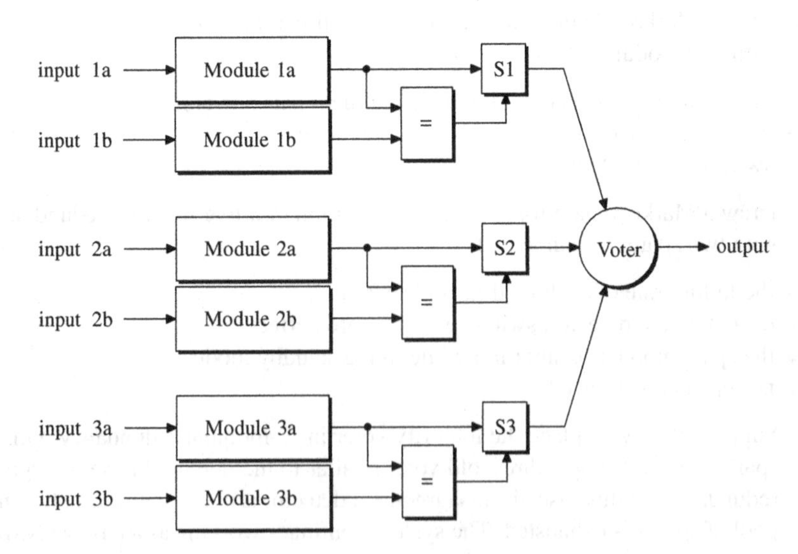

Fig. 4.28 Triplex–duplex redundancy

pair from the system. A threshold voter which is capable of adapting to the decreasing number of inputs is used. When the first duplex pair is removed from voting, its works as a comparator. When the second pair is removed, it simply passes its input signal to the output.
How many module faults can be tolerated by such a configuration?

4.25. Draw a Markov chain for availability evaluation of the triplex–duplex redundancy described in Problem 4.24. Assume that:

- the failure rate of each module is λ and the repair rate is μ,
- the voter is perfect,
- when a pair of modules is switched off from the system due to the failure in one of them, the other module cannot fail while the pair is removed from the system.

References

1. Beiu, V., Quintana, J.M., Avedillo, M.J.: VLSI implementations of threshold logic—a comprehensive survey. IEEE Trans. Neural Netw. **14**(5), 1217–1243 (2003)
2. Bilstein, R.E.: Stages to Saturn: A Technological History of the Apollo/Saturn Launch Vehicle. DIANE Publishing, Darby (1999)
3. COMPAQ: HP/COMPAQ non-stop Himalaya products (2005). http://www.disastertolerance.com/Himalaya.htm
4. Dickinson, M., Jackson, J., Randa, G.: Saturn V launch vehicle digital computer and data adapter. In: Proceedings of the Fall Joint Computer Conference, pp. 501–516 (1964)
5. Ferrara, L.A.: Summary description of the AAP Apollo telescope mount. Technical Report TM-68-1022-3, National Aeronautics and Space Administration (NASA) (1968)
6. Johnson, B.W.: The Design and Analysis of Fault Tolerant Digital Systems. Addison-Wesley, Reading (1989)
7. Losq, J.: Influence of fault-detection and switching mechanisms on the reliability of standby systems. In: Digest 5th International Symposium on Fault-Tolerant Computing, pp. 81–86 (1975)
8. Losq, J.: A highly efficient redundancy scheme: Self-purging redundancy. IEEE Trans. Comput. **25**(6), 569–578 (1976)
9. McAllister, D., Vouk, M.A.: Fault-tolerant software reliability engineering. In: Lyu, M.R. (ed.) Handbook of Software Reliability, pp. 567–614. McGraw-Hill, New York (1996)
10. Moore, E., Shannon, C.: Reliable circuits using less reliable relays. J. Frankl. Inst. **262**(3), 191–208 (1956)
11. Parhami, B.: Voting algorithms. IEEE Trans. Reliab. **43**, 617–629 (1994)
12. Pratt, B., Caffrey, M., Graham, P., Morgan, K., Wirthlin, M.: Improving FPGA design robustness with partial TMR. In: Proceedings of 44th Reliability Physics Symposium, pp. 226–232 (2006)
13. Siewiorek, D.P., McCluskey, E.J.: An iterative cell switch design for hybrid redundancy. IEEE Trans. Comput. **22**(3), 290–297 (1973)
14. Siewiorek, D.P., Swarz, R.S.: Reliable Computer Systems: Design and Evaluation, 3rd edn. A K Peters Ltd, Wellesley (1998)
15. Smith, D.J.: Reliability Engineering. Barnes & Noble Books, New York (1972)

16. von Neumann, J.: Probabilistic logics and synthesis of reliable organisms from unreliable components. In: Shannon, C., McCarthy, J.: (eds.) Automata Studies, pp. 43–98. Princeton University Press, Princeton (1956)
17. Yeh, Y.: Triple-triple redundant 777 primary flight computer. In: Aerospace Applications Conference, Proceedings, 1996 IEEE, vol. 1, pp. 293–307 (1996)

Chapter 5
Information Redundancy

"If you make a mistake and do not correct it, this is called a mistake."

Confucius

This chapter describes how fault tolerance can be achieved by means of coding. Coding is a powerful technique which helps us avoid unwanted information changes during data storage or transmission.

Code selection for a given application is usually guided by the types and rate of errors required to be tolerated, their consequences, and the overhead associated with implementing encoding and decoding. For example, applications susceptible to scratches, such a compact disks or DVDs, use powerful Reed–Solomon codes designed specifically for burst errors. In main memories which store a system's critical, nonrecoverable files, error-correcting Hamming codes are usually employed, whereas the cheaper error-detecting parity codes are used in caches, whose content can be retrieved if found faulty.

The chapter is organized as follows. After a brief survey of the history of coding theory, we define fundamental notions such as code, encoding and decoding, and information rate. Then, we consider the simplest family of codes—parity codes. Afterwards, we present the linear codes. The theory behind linear code provides us with a general framework for constructing many important codes, including Hamming, Cyclic Redundancy Check, and Reed–Solomon. Therefore, we dedicate a large part of this chapter to linear codes. At the end, we cover two more interesting families of codes—unordered codes and arithmetic codes.

5.1 History

Coding theory was originated by two seminal works by Claude Shannon [25] and Richard Hamming [10]. Hamming, who worked at Bell Laboratories in the USA, was studying possibilities for protecting storage devices from the corruption of single bits

E. Dubrova, *Fault-Tolerant Design*, DOI: 10.1007/978-1-4614-2113-9_5,

by using encoding. At that time, only the triplication code, obtained by repeating the data three times, was known to be capable of performing such a task. Hamming was looking for a possibility of creating an error-correcting code with less redundancy. He discovered that such a code should consist of words which differ in a sufficiently large number of bit positions. Hamming introduced the notion of *distance* between two words which is now called in his honor the *Hamming distance*. He constructed a single-error correcting code using only three check bits per four bits of data. Thus, the number of redundant bits in his code was 1.56 times smaller than the 2 out of 3 redundant bits in the triplication code. Hamming published his results in 1950.

Two years before Hamming's publication, Shannon, also at Bell Labs, developed a mathematical theory of communication [25]. He studied how a sender can efficiently communicate over a channel with a receiver. He considered two types of channels: noiseless and noisy. In the former case, the goal is to compress a message in order to minimize the total number of transmitted symbols while allowing the receiver to recover the message. In the latter case, the goal is to add some redundancy to a message so that, in spite of errors caused by a noise, the receiver can still recover the message. Shannon's work showed that the *rate* at which one could communicate over both types of channel is determined by the same underlying rules. Shannon's approach involved encoding messages using long random strings, relying on the fact that long random messages tend to be far away from each other. Shannon has shown that it is possible to encode messages in such a way that the number of required redundant bits is minimized.

Shannon's and Hamming's works made apparent the value of error-correcting codes for the transmission and storage of digital information. A wide variety of efficient codes have been constructed since then, having strong error detecting or correcting capabilities, high information rates, and simple encoding and decoding mechanisms.

5.2 Fundamental Notions

In this section, we introduce the basic notions of coding theory. We assume that our data are in the form of strings of binary bits, 0 or 1. We also assume that the errors occur randomly and independently from each other, but at a predictable overall rate.

5.2.1 Code

A *binary code of length n* is a set of binary n-tuples satisfying some well-defined set of rules. For example, an even parity code contains all n-tuples that have an even number of 1 s. The set $\{0, 1\}^n$ of all possible 2^n binary n-tuples is called a *codespace*.

A *codeword* is an element of the codespace satisfying the rules of the code. To make error detection and error correction possible, a code should contain a nonempty subset of all possible binary n-tuples. For example, a parity code of length n contains 2^{n-1} codewords. An n-tuple not satisfying the rules of the code is called a *word*.

The number of codewords in a code C is called the *size* of C, denoted by $|C|$.

5.2.2 Encoding

Encoding is the process of computing a codeword for a given data. An encoder takes a binary k-tuple representing the data and converts it to a codeword of length n using the rules of the code. The difference $n - k$ between the length of the codeword n and the length of the data k gives us the number of *check bits* which are required to make the encoding.

For example, to encode data into an even parity code, the number of 1's in the data is first calculated. If this number is odd, 1 is appended to the end of the k-tuple. Otherwise, 0 is appended. A parity code uses one check bit.

Separable code is the code in which the check bits can be clearly separated from the data. Parity code is an example of a separable code. *Nonseparable code* is a code in which the check bits cannot be separated from the data. Cyclic code is an example of a nonseparable code.

5.2.3 Information Rate

To encode binary k-bit data, we need a code C consisting of at least 2^k codewords, since any data word should be assigned its own codeword from C. Conversely, a code of size $|C|$ encodes the data of length $k \leq \lceil \log_2 |C| \rceil$ bits. The ratio k/n is called the *information rate* of the code. The information rate determines the redundancy of the code. For example, a triplication code obtained by repeating the data three times has the information rate $1/3$. Only one out of three bits carries the message; the other two bits are redundant.

5.2.4 Decoding

Decoding is the process of extracting data from a given codeword. A decoder reads a codeword and attempts to restore the original data using the rules of the code. For example, if there is no error, then the decoder for a separable code simply truncates the codeword by the number of check bits.

If an error has occurred, it has to be corrected before the data can be retrieved.

Suppose that an error has occurred and a word not satisfying the rules of the code is received by a decoder. A usual assumption in coding theory is that an error involving a few bits is more likely to occur than an error involving many bits [21]. Therefore, the received word is decoded into a codeword which differs from the received word in the fewest positions, if such a codeword is unique. Such a technique is called *maximum likelihood decoding*. The Hamming distance, defined next, is used to measure the number of positions in which two words differ.

5.2.5 Hamming Distance

The *Hamming distance* between two binary n-tuples, x and y, denoted by $H_d(x, y)$, is the number of bit positions in which the n-tuples differ. For example, 4-tuples $x = [0011]$ and $y = [0101]$ differ in two bit positions, so $H_d(x, y) = 2$. Hamming distance gives us an estimate of how many bit errors have to occur to change x into y.

Hamming distance is a genuine metric on the codespace $\{0, 1\}^n$. A *metric* is a function which associates any two objects in a set with a number [5]. It has the following properties:

1. **Reflexivity**: $H_d(x, y) = 0$ if and only if $x = y$.
2. **Symmetry**: $H_d(x, y) = H_d(y, x)$.
3. **Triangle inequality**: $H_d(x, y) + H_d(y, z) \geq H_d(x, z)$.

The metric properties of the Hamming distance allow us to use the geometry of the codespace to reason about the codes. As an example, consider the codespace $\{0, 1\}^3$ illustrated by a cube shown in Fig. 5.1. The codewords $\{000, 011, 101, 110\}$ are marked with large solid dots. It is easy to see that the Hamming distance satisfies the metric properties listed above, e.g. $H_d(000, 011) + H_d(011, 111) = 2 + 1 = 3 = H_d(000, 111)$.

5.2.6 Code Distance

The *code distance* of a code C, denoted by C_d, is the minimum Hamming distance between any two distinct pairs of codewords of C. The code distance determines the error-detecting and error-correcting capabilities of a code. For example, consider the code $\{000, 011, 101, 110\}$ in Fig. 5.1. The code distance of this code is 2. Any

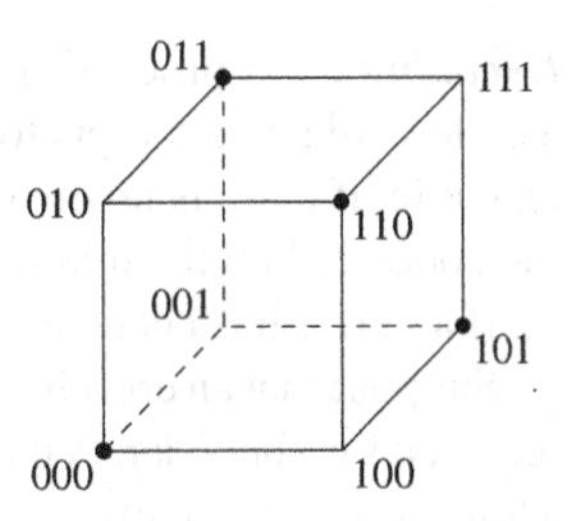

Fig. 5.1 Code $\{000, 011, 101, 110\}$ in the codespace $\{0, 1\}^3$

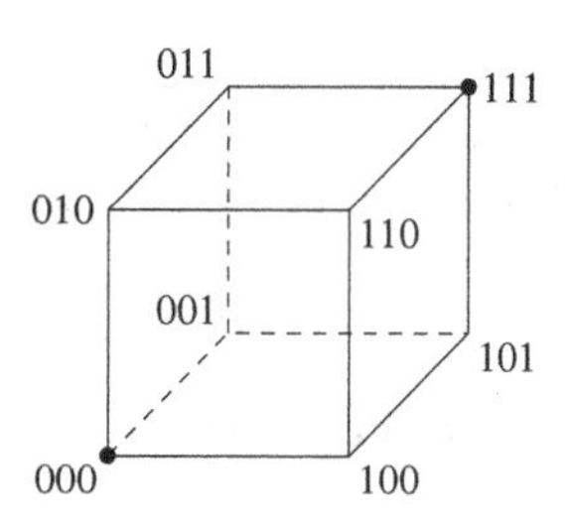

Fig. 5.2 Code {000, 111} in the codespace $\{0, 1\}^3$

single-bit error in any codeword produces a word which has a distance of 1 from the affected codeword. Since all codewords have a distance of 2 from each other, any single-bit error is detected.

As another example, consider the code {000, 111} shown in Fig. 5.2. Suppose an error has occurred in the first bit of the codeword [000]. The resulting word [100] has a distance of 1 from the codeword [000] and a distance of 2 from the codeword [111]. Using the maximum likelihood decoding rule, we correct [100] to the codeword [000], which is the closest.

The code {000, 111} is a triplication code, obtained by repeating the data three times. It can be viewed as a TMR implemented in the information domain. In TMR, the voter compares the output values of the modules. In a triplication code, a decoder analyzes the bits of the received words. In both cases, the majority of values of bits can be used to decide the correct value.

In general, if a code is used only for error detection, then, to detect all errors in d or fewer bits, it should have the code distance $C_d \geq d + 1$ [21]. This is because if $C_d \geq d + 1$, no errors in d or fewer bits can change one codeword into another codeword. Conversely, if $C_d < d + 1$, then there exist a pair of codewords on a Hamming distance d or less. Thus, an error in d or fewer bits can change one codeword into another. For example, the duplication code {00, 11}, which has the code distance 2, cannot detect double-bit errors because a double-bit error would change [00] into [11], or vice versa.

Similarly, a code can correct all errors in c or fewer bits if and only if its code distance is $C_d \geq 2c + 1$ [21]. Then, any word with c^*-bit error, $c^* \leq c$, differs from the fault-free original codeword in c^* bits and from any other codeword in at least $2c + 1 - c^* > c^*$ bits. Conversely, if $C_d < 2c + 1$, then a c^*-bit error, $c^* \leq c$, may result in a word which is at least as close to another codeword as it is to the fault-free original codeword. For example, the duplication code {00, 11}, which has the code distance 2, cannot correct single-bit errors because any single-bit change results in a word which is equally close to both codewords, [00] and [11]. For instance, an error in the first bit in [00] results in the word [10]. Since $H_d(00, 01) = H_d(01, 11)$, the error cannot be corrected. On the other hand, the triplication code {000, 111} can correct single-bit errors. We can use a decoder which votes on the input bits and outputs the majority as the decoded data.

Finally, it is possible to construct a decoder which corrects all errors in c or fewer bits and simultaneously detects up to a additional errors if and only if the code distance is $C_d \geq 2c + a + 1$ [21]. For example, a code with the code distance 4 can correct single-bit errors and simultaneously detect one more additional error. We leave the proof of this statement as an exercise for the reader (Problem 5.7).

5.3 Parity Codes

Parity codes are the oldest family of codes. They were used to detect errors in the calculations of the relay-based computers in the late 1940s [10].

5.3.1 Definition and Properties

The *even (odd) parity code* of length n consists of all binary n-tuples that contain an even (odd) number of 1's. Typically, the first $n - 1$ bits of a codeword are data carrying the information, while the last bit is the check bit, determining the parity of the codeword. The information rate of a parity code of length n is $(n - 1)/n$.

For example, if the data [0011] are encoded using an even parity code, then the resulting codeword is [00110]. Since the data contain an even number of 1's, the appended check bit is 0.

On the other hand, if the data [0011] are encoded using an odd parity code, then the resulting codeword is [00111]. The check bit 1 makes the number of 1's in the codeword odd.

A parity code of length n contains one half of all possible 2^n binary n-tuples. Therefore, it has the size 2^{n-1}. The code distance of a parity code of any length is 2. To show that this is true, let us consider the set of $n - 1$-tuples representing the binary data encoded by a parity code of length n. There are at least two $n - 1$-tuples on Hamming distance 1. One of them has an odd number of 1's and another one has an even number of 1's. For an even (odd) parity code, the check bit for the former data word is 1 (0), while the check bit for the latter data word is 0 (1). The difference between check bits makes the Hamming distance between the resulting codewords 2. Thus, the parity code of length n has the code distance 2, for any $n > 1$.

A parity code can detect single-bit errors and multiple-bit errors which affect an odd number of bits. If an odd number of bits is affected, then the number of 1's in a codeword changes from even to odd (for an even parity code) or from odd to even (for an odd parity code). For example, consider the codeword [00111] of a 5-bit odd parity code. If an error affects one bit, say the first, we get the word [10111] which has an even number of 1's. So, we detect that this word is not a codeword. If an error affects three bits, say the first three, we get the word [11011] which has an even number of 1's. Again, we detect the error. On the other hand, if an error affects an even number of bits, say the first two, we get the word [11111] which has an odd number of 1's. Since this word is a codeword, the error is not detected. A parity code cannot detect any multiple-bit errors which affect an even number of bits.

A parity code can only detect errors, but it cannot correct them, since the position of the erroneous bit(s) is not possible to locate. For example, suppose that we use an even parity code to protect the transmission of 5-bit data. If we receive the word [110100], then we know that an error has occurred. However, we do not know which codeword has been sent. If we assume that a single-bit error has occurred during the transmission, then it is equally likely that one of the six codewords [010100], [100100], [111100], [110000], [110110], or [110101] has been sent.

Example 5.1. Construct the odd parity code for 3-bit data.

Eight 3-bit words which we need to encode are shown at the left-hand side of Table 5.1. We append to each 3-bit word $d = [d_0 d_1 d_2]$ one parity bit which makes the overall number of 1's in the resulting codeword $c = [c_0 c_1 c_2 c_3]$ odd. As a result, we get the codewords listed at the right-hand side of Table 5.1.

5.3.2 Applications

Due to their simplicity, parity codes are used in many applications where an operation can be repeated in the case of an error, or where simply detecting the error is helpful.

For example, many processors use parity to protect their data and instruction caches (both storage and tag arrays) [12]. Since the cache data are just a copy of main memory, it can be discharged and retrieved if it is found faulty. For data cache storage array, one parity bit per byte (8 bits) of data is usually computed. For data and instruction cache tag arrays, one parity bit per entry is typically computed.

A general diagram of a memory protected by a parity code is shown in Fig. 5.3 Before being written into a memory, the data are encoded by computing its parity. The generation of check bits is done by a Parity Generator (PG) implemented as a tree of XOR gates. Figure 5.4 shows a logic diagram of an even parity generator for 4-bit data $[d_0 d_1 d_2 d_3]$.

When data are written into a memory, check bits are written along with the corresponding bytes of data. For example, for systems with a 32-bit wide memory data

Table 5.1 The defining table for the odd parity code for 3-bit data

Data			Codeword			
d_0	d_1	d_2	c_0	c_1	c_2	c_3
0	0	0	0	0	0	1
0	0	1	0	0	1	0
0	1	0	0	1	0	0
0	1	1	0	1	1	1
1	0	0	1	0	0	0
1	0	1	1	0	1	1
1	1	0	1	1	0	1
1	1	1	1	1	1	0

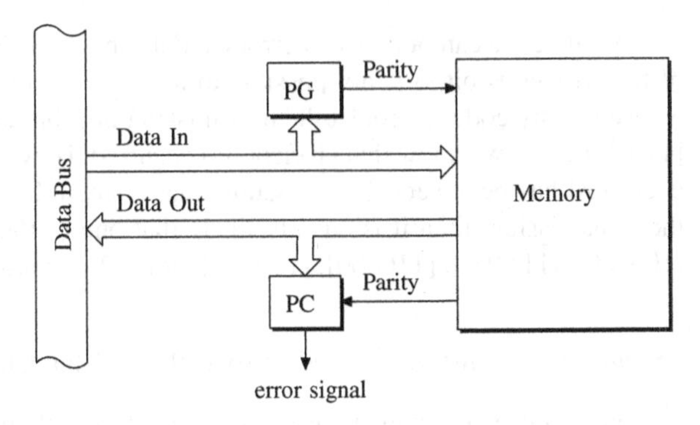

Fig. 5.3 A memory protected by a parity code; PG = Parity Generator; PC = Parity Checker

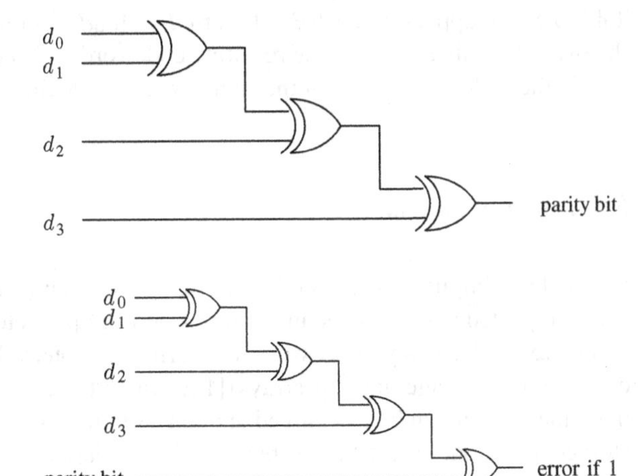

Fig. 5.4 Logic diagram of a even parity generator for 4-bit data $[d_0d_1d_2d_3]$

Fig. 5.5 Logic diagram of a even parity checker for 4-bit data $[d_0d_1d_2d_3]$

path, four check bits are attached to the data and the resulting 36-bit codeword is stored in the memory. Some systems, like the Pentium processor, have a 64-bit wide memory data path [1]. In such a case, eight check bits are attached to the data. The resulting codeword is 72 bits long.

When the data are read back from a memory, check bits are recomputed and compared to the previously stored check bits. Re-computation and comparison is done by a Parity Checker (PC). Figure 5.5 shows a logic diagram of an even parity checker for 4-bit data $[d_0d_1d_2d_3]$. The logic diagram is similar to that of a parity generator, except that one more XOR gate is added to compare the recomputed check bit to the previously stored check bit. If the check bits disagree, the output of the XOR gate is 1. Otherwise, the output is 0.

Any recomputed check bit that does not match the stored check bit indicates that there is at least one bit error in the corresponding byte of data, or in the parity bit itself. In such a case, an error signal is sent to the central processing unit (CPU) to indicate that the memory data are not valid and to instruct the system to halt.

All operations related to the error detection (encoding, decoding, comparison) are done by the memory control logic on the motherboard, in the chip, or, for some systems, in the CPU. The memory itself only stores the parity bits, just as it stores the data bits. The check-bit generation and checking is done in parallel with the writing and reading of the memory using the logic which is much faster that the memory itself. Therefore, these operations do not slow down the operation of the memory. The system is not slowed down either, because it does not wait for a "no error" signal from the parity checker to perform its actions. It halts only if an error is detected.

For example, suppose the data which is written in the memory are [0110110] and an odd-parity code is used. Then, the check bit 1 is stored along with the data to make the overall parity odd. Suppose that the codeword which is read out of the memory is [01111101]. The recomputed check bit is 0. Since the recomputed check bit disagrees with the stored one, we know that an error has occurred. However, we do not know which bit is changed.

Parity codes are also commonly used for detecting transmission errors in the external address and data buses. Typically, there is one parity bit for each byte of the data bus and one parity bit for the address bus [12].

5.3.3 Horizontal and Vertical Parity Codes

A modification of a parity code is the *horizontal and vertical parity code*, which arranges the data in a 2-dimensional array and adds one check bit to each row and one check bit to each column. Since each row and column intersects in only one bit position, such a technique enables single bit errors within a block of data to be corrected.

As an example, consider the horizontal and vertical odd parity code shown in Table 5.2. The check bits are highlighted in bold. Any single-bit error causes the values of check bits in one row and one column to change. By intersecting these rows and columns, we can find the location of the error.

Similarly to the parity code, the horizontal and vertically parity code detects multiple-bit errors affecting an odd number of bits. Also, it detects any two-bit errors. However, it may miss detecting an error affecting an even number of bits if this number is larger than two. For example, a 4-bit error affecting the first and second bits of the first and second rows of the code in Table 5.2 will not be detected.

The horizontal and vertical parity code is an example of *overlapping parity* codes, in which each bit of data is covered by more than one check bit. Another famous overlapping parity code is Hamming code. We will describe this code in the next section, after introducing the theory of linear codes.

Table 5.2 An example of the horizontal and vertical odd parity code for 4 × 4-bit data

0	1	0	1	**1**
1	0	1	1	**0**
0	0	1	0	**0**
1	0	0	0	**0**
1	**0**	**1**	**1**	**0**

5.4 Linear Codes

Linear codes provide us with a general framework for constructing many important codes, including the Hamming code. The discussion of linear codes requires some knowledge of finite fields and linear algebra, which we briefly review below. For a more detailed treatment of this topic, the reader is referred to [17] and [28].

5.4.1 Basic Notions

Finite Field $\mathbb{F}_2$

A *finite field* $\mathbb{F}_2$ is the set $\{0, 1\}$ together with two binary operations, addition "$\oplus$" and multiplication "$\cdot$", such that the following properties are satisfied for all $a, b, c \in \mathbb{F}_2$:

1. $a \cdot b \in \mathbb{F}_2$ and $a \oplus b \in \mathbb{F}_2$.
2. $a \oplus (b \oplus c) = (a \oplus b) \oplus c$.
3. $a \oplus b = b \oplus a$
4. There exists an element $\mathbf{0} \in \mathbb{F}_2$ such that $a \oplus \mathbf{0} = a$.
5. For each $a \in \mathbb{F}_2$, there exists an element $-a \in \mathbb{F}_2$ such that $a \oplus (-a) = \mathbf{0}$.
6. $a \cdot (b \cdot c) = (a \cdot b) \cdot c$.
7. $a \cdot b = b \cdot a$.
8. There exists an element $\mathbf{1} \in \mathbb{F}_2$ such that $a \cdot 1 = 1 \cdot a = a$.
9. For each $a \in \mathbb{F}_2$, such that $a \neq \mathbf{0}$, there exists an element $a^{-1} \in \mathbb{F}_2$ such that $a \cdot a^{-1} = \mathbf{1}$.
10. $a \cdot (b \oplus c) = a \cdot b \oplus a \cdot c$.

The above properties are satisfied if

- addition "$\oplus$" is defined as the addition modulo 2,
- multiplication "$\cdot$" is defined as the multiplication modulo 2,
- $\mathbf{0} = 0$ and $\mathbf{1} = 1$.

We leave the proof of this statement as an exercise for the reader. Throughout the chapter, we assume this definition of $\mathbb{F}_2$. Note that the addition modulo 2 is equivalent to the XOR and the multiplication modulo 2 is equivalent to the AND.

Vector space V^n

Let $\mathbb{F}_2^n$ be the set of all n-tuples containing elements from $\mathbb{F}_2$. For example, for $n = 3$, $\mathbb{F}_2^3 = \{000, 001, 010, 011, 100, 110, 111\}$.

A *vector space* V^n over a field $\mathbb{F}_2$ is a subset of $\mathbb{F}_2^n$ together with two binary operations, "$\oplus$" and "$\cdot$", which satisfy the axioms listed below. Elements of V^n are called *vectors*. Elements of $\mathbb{F}_2$ are called *scalars*. The operation "$\oplus$" is called *vector addition*. It takes two vectors $v, u \in V^n$ and maps them into a third vector which is the sum $v \oplus u$ of these two vectors. The operation "$\cdot$" is called *scalar multiplication*. It takes a scalar $a \in \mathbb{F}_2$ and a vector $v \in V^n$ and rescales v by a, resulting in a vector $a \cdot v$.

To be a vector space, the set V^n and the operations of addition and multiplication should satisfy the following axioms for all vectors $v, u, w \in V^n$ and all scalars $a, b, c \in \mathbb{F}_2$:

1. $v \oplus u \in V^n$.
2. $v \oplus u = u \oplus v$.
3. $v \oplus (u \oplus w) = (v \oplus u) \oplus w$.
4. There exists an element $\mathbf{0} \in V^n$ such that $v \oplus \mathbf{0} = v$.
5. For each $v \in V^n$, there exists an element $-v \in V^n$ such that $v \oplus (-v) = \mathbf{0}$.
6. $a \cdot v \in V^n$.
7. There exists an element $\mathbf{1} \in \mathbb{F}_2$ such that $\mathbf{1} \cdot v = v$.
8. $a \cdot (v \oplus u) = (a \cdot v) \oplus (a \cdot u)$.
9. $(a \oplus b) \cdot v = a \cdot v \oplus b \cdot v$.
10. $(a \cdot b) \cdot v = a \cdot (b \cdot v)$.

The above axioms are satisfied if

- vector addition "$\oplus$" is defined as the element-wise addition modulo 2,
- scalar multiplication "$\cdot$" is defined as the element-wise multiplication modulo 2,
- $\mathbf{0}$ is the zero vector, containing all 0's.
- $\mathbf{1}$ is the one vector, containing all 1's.

"Element-wise addition" means that, in order to add two n-bit vectors v and u, we add the ith element of v to the ith element of u, for all $i \in \{0, 1, \dots, n-1\}$. For example, if $v = [0110]$ and $u = [0011]$, then $v \oplus u = [0101]$.

Similarly, "element-wise multiplication" means that, in order to multiply an n-bit vector v by a scalar a, we multiply the ith element of v by a, for all $i \in \{0, 1, \dots, n-1\}$. Since in $\mathbb{F}_2$ there are only two scalars, 0 and 1, for any vector v, there are only two possible outcomes of scaling: $1 \cdot v = v$ and $0 \cdot v = 0$ (note that the left-hand side zero is a scalar, while the right-hand side zero is a vector).

Throughout the chapter, we assume the above definition of V^n.

Subspace

A *subspace* is a subset of a vector space that is itself a vector space.

Span

A set of vectors $\{v_0, \dots, v_{k-1}\}$ is said to *span* a vector space V^n if any $v \in V^n$ can be written as $v = a_0 v_0 \oplus a_1 v_1 \oplus \dots \oplus a_{k-1} v_{k-1}$, where $a_0, \dots, a_{k-1} \in \mathbb{F}_2$.

For example, the set of vectors $\{1000, 0100, 0010, 0001\}$ spans the vector space V^4 because any other vector in V^4 can be obtained as a linear combination of these vectors.

Linear independence

A set of vectors $\{v_0, \dots, v_{k-1}\}$ of V^n is *linearly independent* if $a_0 v_0 \oplus a_1 v_1 \oplus \dots \oplus a_{k-1} v_{k-1} = 0$ implies $a_0 = a_1 = \dots = a_{k-1} = 0$. In other words, none of the vectors in the set can be written as a linear combination of finitely many other vectors in the set.

Checking linear independence of vectors is particularly easy for sets of a small size. It is worth considering the cases of $k = 1, 2$ and 3 which we will use most for constructing linear codes throughout the section.

For $k = 1$, a single vector is linearly independent if and only if it is not the zero vector.

For $k = 2$, a pair of vectors is linearly independent if and only if: (1) neither of them is the zero vector, and (2) they are not equal to each other.

For $k = 3$, three vectors are linearly independent if and only if: (1) neither of them is the zero vector, (2) no vector repeats twice, and (3) no two of them sum up to the third.

Basis

A *basis* is a set of vectors in a vector space V^n that are linearly independent and span V^n.

For example, the set of vectors $\{100, 010, 001\}$ is a basis for the vector space V^3, because they are linearly independent and any other vector in V^3 can be obtained as a linear combination of these vectors.

Dimension

The *dimension* of a vector space is the number of vectors in its basis.

For example, the dimension of the vector space V^3 is 3.

5.4.2 Definition

A (n, k) linear code over the field $\mathbb{F}_2$ is a k-dimensional subspace of V_n. In other words, an (n, k) linear code is a subspace of V^n which is spanned by k linearly independent vectors.

All codewords of an (n, k) linear code can be written as a linear combination of the k basis vectors $\{v_0, \ldots, v_{k-1}\}$ as follows:

$$c = d_0 v_0 \oplus d_1 v_1 \oplus \ldots \oplus d_{k-1} v_{k-1}. \tag{5.1}$$

Since a different codeword is obtained for each different combination of coefficients $d_0, d_1, \ldots, d_{k-1}$, the Eq. (5.1) gives us a systematic method to encode k-bit data $d = [d_0 d_1 \ldots d_{k-1}]$ into an n-bit codeword $c = [c_0 c_1 \ldots c_{n-1}]$.

The information rate of an (n, k) linear code is k/n.

Example 5.2. Construct a (4, 2) linear code.

Since $k = 2$, the data we are encoding are 2-bit words $\{00, 01, 10, 11\}$. We should encode them, so that the resulting 4-bit codewords form a two-dimensional subspace of V^4. To achieve this, we need to select two linearly independent vectors as a basis of the two-dimensional subspace. One possibility is to choose $v_0 = [1000]$ and $v_1 = [0110]$. These vectors are linearly independent since neither of them is the zero vector and they are not equal to each other.

To encode the data $d = [d_0 d_1]$ into the codeword $c = [c_0 c_1 c_2 c_3]$, we compute the linear combination of the basis vectors v_0 and v_1 as $c = d_0 v_0 \oplus d_1 v_1$. As a result, we get:

$$\begin{aligned} d = [00] &: c = 0 \cdot [1000] \oplus 0 \cdot [0110] = [0000] \\ d = [01] &: c = 0 \cdot [1000] \oplus 1 \cdot [0110] = [0110] \\ d = [10] &: c = 1 \cdot [1000] \oplus 0 \cdot [0110] = [1000] \\ d = [11] &: c = 1 \cdot [1000] \oplus 1 \cdot [0110] = [1110]. \end{aligned}$$

Recall that the addition is modulo 2, so $1 \oplus 1 = 0$.

5.4.3 Generator Matrix

The encoding we performed using the Eq. (5.1) can be formalized by introducing the *generator matrix* G, whose rows are the basis vectors v_0 through v_{k-1}:

$$G = \begin{bmatrix} v_0 \\ v_1 \\ \dots \\ v_{k-1} \end{bmatrix}.$$

For instance, the generator matrix for the basis vectors v_0 and v_1 from Example 5.2 is

$$G = \begin{bmatrix} 1\,0\,0\,0 \\ 0\,1\,1\,0 \end{bmatrix}. \tag{5.2}$$

Now we can compute a codeword c by multiplying the data vector d by the generator matrix G:

$$c = dG.$$

Note that in Example 5.2, the first two bits of each codeword are exactly the same as the data bits, i.e., the code we have constructed is a separable code. Separability is a desirable feature since, after correction, the data can be easily retrieved by truncating the last $n-k$ bits of a codeword. We can construct a separable linear code by choosing basis vectors which result in a generating matrix of the form $[I_k A]$, where I_k is an identity matrix of size $k \times k$.

Note also that the code distance of the code in Example 5.2 is 1. Such a code is not useful, since it cannot detect any errors. Code distance is related to the properties of the basis vectors. Consider the vector $v_1 = [1000]$ from Example 5.2. Since it is a basis vector, it is a codeword. Any code is a subspace V^n. Thus, a code is itself a vector space. A vector space is closed under addition (Axiom 1). Therefore, for any codeword c, the word $c \oplus 1000$ also belongs to the code. Since the Hamming distance between c and $c \oplus 1000$ is 1, the code distance of the code is 1. Consequently, if we want to construct a code with a code distance C_d, we need to choose basis vectors with the number of 1's greater than or equal to C_d.

Example 5.3. List all codewords of the (6, 3) linear code constructed using the basis vectors $v_0 = [100011]$, $v_1 = [010110]$ and $v_2 = [001101]$.

The generator matrix for this code is

$$G = \begin{bmatrix} 1\,0\,0\,0\,1\,1 \\ 0\,1\,0\,1\,1\,0 \\ 0\,0\,1\,1\,0\,1 \end{bmatrix}. \tag{5.3}$$

Note that G is in the form $[I_3 A]$, where I_3 is an identity matrix of size 3×3, so the resulting code is separable.

To encode 3-bit data $d = [d_0 d_1 d_2]$ into a codeword $c = [c_0 c_1 c_2 c_3 c_4 c_5]$, we multiply d by G. For example, the data $d = [011]$ is encoded into the codeword

$$c = 0 \cdot [100011] \oplus 1 \cdot [010110] \oplus 1 \cdot [001101] = [011011].$$

Similarly, we can encode all other 3-bit data words. The resulting code is shown in Table 5.3.

Linear codes give us a general framework for constructing many codes. For example, using generator matrices, we can construct triplication codes and parity codes. The triplication code for single-bit data has the following generator matrix:

$$G = [111].$$

The even parity code for 4-bit data has the following generator matrix:

$$G = \begin{bmatrix} 1\,0\,0\,0\,1 \\ 0\,1\,0\,0\,1 \\ 0\,0\,1\,0\,1 \\ 0\,0\,0\,1\,1 \end{bmatrix}.$$

As we can see, both these codes are linear codes.

Table 5.3 Defining table for a (6, 3) linear code

Data			Codeword					
d_0	d_1	d_2	c_0	c_1	c_2	c_3	c_4	c_5
0	0	0	0	0	0	0	0	0
0	0	1	0	0	1	1	0	1
0	1	0	0	1	0	1	1	0
0	1	1	0	1	1	0	1	1
1	0	0	1	0	0	0	1	1
1	0	1	1	0	1	1	1	0
1	1	0	1	1	0	1	0	1
1	1	1	1	1	1	0	0	0

5.4.4 Parity Check Matrix

To detect errors in an (n, k) linear code, we use an $(n - k) \times n$ matrix H, called the *parity check matrix* of the code. This matrix has the property that, for any codeword c:

$$H \cdot c^T = 0.$$

where c^T denotes a transpose of c. Recall that the transpose of an $n \times k$ matrix A is the $k \times n$ matrix obtained by defining the ith column of A^T to be the ith row of A.

The parity check matrix is related to the generator matrix by the equation

$$HG^T = 0.$$

This equation implies that, if data d are encoded to a codeword dG using the generator matrix G, then the product of the parity check matrix and the encoded data should be zero. This is true because

$$H(dG)^T = H(G^T d^T) = (HG^T)d^T = 0d^T = 0.$$

If a generator matrix is of the form $G = [I_k A]$, then a parity check matrix is of the form

$$H = [A^T I_{n-k}].$$

This can be proved as follows:

$$HG^T = A^T I_k \oplus I_{n-k} A^T = A^T \oplus A^T = 0.$$

Example 5.4. Construct the parity check matrix H for the generator matrix G given by (5.3).

Since the generator matrix G is of the form $[I_3 A]$, where

$$A = \begin{bmatrix} 0\,1\,1 \\ 1\,1\,0 \\ 1\,0\,1 \end{bmatrix},$$

we get the following the parity check matrix:

$$H = [A^T I_3] = \begin{bmatrix} 0\,1\,1\,1\,0\,0 \\ 1\,1\,0\,0\,1\,0 \\ 1\,0\,1\,0\,0\,1 \end{bmatrix}. \tag{5.4}$$

5.4.5 Syndrome

From the discussion in the previous section, we can conclude that encoded data can be checked for errors by multiplying it by the parity check matrix:

$$s = Hc^T. \tag{5.5}$$

The resulting k-bit vector s is called the *syndrome*. If the syndrome is zero, no error has occurred. If s matches one of the columns of H, then a single-bit error has occurred. The position of the matching column in H corresponds to the bit position of the error, so the error can be located and corrected. For example, if the syndrome coincides with the second column of H, the error is in the second bit of the codeword. If the syndrome is not zero and it is not equal to any of the columns of H, then a multiple-bit error has occurred. The location of a multiple-bit error is unknown, so it cannot be corrected.

As an example, consider the data $d = [110]$ encoded using the (6, 3) linear code from Example 5.3 as $c = dG = [110101]$. Suppose that an error occurs in the second bit of c, transforming it to [100101]. By multiplying this word by the parity check matrix (5.4), we obtain the syndrome $s = [110]$. The syndrome matches the second column of the parity check matrix H, indicating that the error has occurred in the second bit.

5.4.6 Construction of Linear Codes

Suppose we would like to construct a linear code with the code distance C_d for k-bit data. It is known that a given code distance can be ensured by selecting a parity check matrix which satisfies the following property [21].

Property 5.1. Let H be a parity check matrix for an (n, k) linear code. Then, this code has the code distance $C_d \geq \Delta$ if and only if every subset of $\Delta - 1$ columns of its parity check matrix H are linearly independent.

For example, in order to construct a linear code with a code distance of at least 3, we should select a parity check matrix in which every pair of columns are linearly independent vectors. Recall that this is equivalent to the requirement of not having a zero column and not having columns repeating twice.

In order to maximize the information rate, we should select the minimum code length n which satisfies the Property 5.1. However, there is no formula which gives us the minimum n for given k and C_d. Only the following bound, known as the *Singleton bound* [21], is available:

$$n \geq C_d + k - 1. \tag{5.6}$$

Table 5.4 Defining table for a (4, 2) linear code

Data		Codeword			
d_0	d_1	c_0	c_1	c_2	c_3
0	0	0	0	0	0
0	1	0	1	1	0
1	0	1	0	1	1
1	1	1	1	0	1

Equation (5.6) is a necessary condition for Property 5.1 to hold, meaning that Property 5.1 never holds if $n < C_d + k - 1$. However, Eq. 5.6 is not a sufficient condition for Property 5.1. Property 5.1 may not hold for some n satisfying $n \geq C_d + k - 1$.

For example, if we construct a linear code with $C_d = 3$ for 3-bit data, then Eq. 5.6 gives us the bound $n \geq 6$. However, as we will show in Example 5.6, it is not possible to construct a (6, 3) linear code satisfying Property 5.1. A code length of at least $n = 7$ is required.

In the parity check matrix (5.2) of the code which we constructed in Example 5.2, the first column is zero:

$$H = \begin{bmatrix} 0\,1\,1\,0 \\ 0\,0\,0\,1 \end{bmatrix}.$$

Therefore, columns of H are linearly dependent and the code distance is 1.

Let us modify H to construct a new code with a code distance of at least 2. We need to replace the zero column by another column. For example, let us use the column containing all 1's. Then, we get the following parity check matrix:

$$H = \begin{bmatrix} 1\,1\,1\,0 \\ 1\,0\,0\,1 \end{bmatrix} = [A^T I_2].$$

So, now A is

$$A = \begin{bmatrix} 1\,1 \\ 1\,0 \end{bmatrix}$$

and therefore the generator matrix G is

$$G = [I_2 A^T] = \begin{bmatrix} 1\,0\,1\,1 \\ 0\,1\,1\,0 \end{bmatrix}.$$

Using this generator matrix, the data words can be encoded as dG resulting in the (4,2) linear code shown in Table 5.4.

By examining the codewords, the reader can verify that the code distance of the resulting (4, 2) code is indeed 2.

Example 5.5. Construct a linear code for single-bit data with a code distance of at least 3.

First, we create a parity check matrix in the form $[A^T I_{n-k}]$ in which every pair of columns is linearly independent. This can be achieved by ensuring that each column is nonzero and no column is repeated twice.

From Eq.(5.6), we can conclude that the minimum codeword length n for which these two conditions are satisfied is $n \geq 3$. Indeed, $n = 2$ is not sufficient, because then H is of size 1×2, so we only have the following two choices:

$$H = [01] \quad \text{or} \quad H = [11].$$

In both cases, two of the columns of H are linearly dependent.

For $n = 3$, we can construct the following matrix H:

$$H = \begin{bmatrix} 1\,1\,0 \\ 1\,0\,1 \end{bmatrix}.$$

We can see that every pair of its columns is linearly independent.

The matrix A^T in this case is

$$A^T = \begin{bmatrix} 1 \\ 1 \end{bmatrix}.$$

So, $A = [11]$ and therefore G is

$$G = \begin{bmatrix} 1\,1\,1 \end{bmatrix}.$$

The resulting (3,1) code is the triplication code {000, 111}.

Example 5.6. Construct a linear code for 3-bit data with a code distance of at least 4.

First, we create a parity check matrix in the form $[A^T I_{n-k}]$ in which every three columns are linearly independent. This can be achieved by ensuring that each column is nonzero, no column is repeated twice, and no two columns sum up to a third column.

From the Eq. (5.6), we can conclude that the minimum codeword length n for which these three conditions are satisfied is $n \geq 6$.

However, $n = 6$ is not sufficient to satisfy the third condition. For any subset of six out of seven nonzero columns of length 3, there are always two columns which sum up to a third column from the subset.

Let us try $n = 7$. Then H is of size 4×7. The last four columns of H form the identity matrix I_4. We need to select three columns for the matrix A^T from the 11 remaining nonzero columns of length 4.

We can use the property that, for any $v, u \in V^n$, if both v and u contain an odd number of 1's and $v \neq u$, then vector $v \oplus u$ contains an even number of 1's which is greater than 0. We leave the derivation of the proof of this property as an exercise for the reader (Problem 5.19).

It follows from this property that, if we form A^T from the columns which have three 1's, then the sum of any two columns of the resulting H will contain two 1's. For example, we can choose the following A^T:

$$A^T = \begin{bmatrix} 1\,1\,1 \\ 1\,1\,0 \\ 1\,0\,1 \\ 0\,1\,1 \end{bmatrix}.$$

Then, the parity check matrix is:

$$H = \begin{bmatrix} 1\,1\,1\,1\,0\,0\,0 \\ 1\,1\,0\,0\,1\,0\,0 \\ 1\,0\,1\,0\,0\,1\,0 \\ 0\,1\,1\,0\,0\,0\,1 \end{bmatrix}.$$

The resulting generator matrix is:

$$G = \begin{bmatrix} 1\,0\,0\,1\,1\,1\,0 \\ 0\,1\,0\,1\,1\,0\,1 \\ 0\,0\,1\,1\,0\,1\,1 \end{bmatrix}.$$

The (7,3) linear code which we constructed is shown in Table 5.5.

5.4.7 Hamming Codes

Hamming codes are an important family of linear codes. They are named after Richard Hamming, who developed the first single-error correcting Hamming code and its extended version, single-error correcting double-error detecting Hamming code, in the early 1950s [10]. These codes remain important even today.

Table 5.5 Defining table for a (7, 3) linear code from Example 5.6

Data			Codeword						
d_0	d_1	d_2	c_0	c_1	c_2	c_3	c_4	c_5	c_6
0	0	0	0	0	0	0	0	0	0
0	0	1	0	0	1	1	0	1	1
0	1	0	0	1	0	1	1	0	1
0	1	1	0	1	1	0	1	1	0
1	0	0	1	0	0	1	1	1	0
1	0	1	1	0	1	0	1	0	1
1	1	0	1	1	0	0	0	1	1
1	1	1	1	1	1	1	0	0	0

Table 5.6 Examples of data lengths for which a Hamming code exists

Data Length	Code Length	Information Rate
1	3	0.333
4	7	0.571
11	15	0.733
26	31	0.839
57	63	0.905
120	127	0.945
247	255	0.969

An (n, k) linear code is a *Hamming code* if its parity check matrix contains $2^{n-k}-1$ columns representing all possible nonzero binary vectors of length $n-k$.

Here is an example of the parity check matrix of a $(7, 4)$ Hamming code:

$$H = \begin{bmatrix} 1\,1\,0\,1\,1\,0\,0 \\ 1\,0\,1\,1\,0\,1\,0 \\ 1\,1\,1\,0\,0\,0\,1 \end{bmatrix}. \tag{5.7}$$

Since the parity check matrix of a Hamming code does not contain zero column and none of its columns repeat twice, every pair of its columns is linearly independent. So, by Property 5.1, the code distance of a Hamming code is 3. Therefore, a Hamming code can correct single-bit errors. The information rate of an (n, k) Hamming code is $k/(2^{n-k}-1)$.

Since the parity check matrix of a Hamming code has $2^{n-k}-1$ columns and the dimension of a parity check matrix is $(n-k) \times n$, the data length k and the code length n of a Hamming code are related as follows:

$$n = 2^{n-k} - 1. \tag{5.8}$$

Hamming codes exist only for those pairs (n, k) which satisfy Eq. 5.8. For any $r > 2$, Eq. 5.8 has an solution for $k = 2^r - r - 1$ and $n = 2^r - 1$. Table 5.6 shows examples of data lengths for which a Hamming code exists. It also shows the information rate of the resulting code. The information rate of Hamming codes is the highest possible for codes with code distance 3 and code length $2^r - 1$.

Example 5.7. Construct the generator matrix corresponding to the parity check matrix (5.7).

Since H is in the standard form $[A^T I_3]$ where

$$A^T = \begin{bmatrix} 1\,1\,0\,1 \\ 1\,0\,1\,1 \\ 1\,1\,1\,0 \end{bmatrix},$$

the generator matrix is also in the standard form $G = [I_4 A]$:

$$G = \begin{bmatrix} 1\,0\,0\,0\,1\,1\,1 \\ 0\,1\,0\,0\,1\,0\,1 \\ 0\,0\,1\,0\,0\,1\,1 \\ 0\,0\,0\,1\,1\,1\,0 \end{bmatrix}.$$

Suppose the data to be encoded are $d = [1110]$. We multiply d by G to get the codeword $c = [1110001]$. Suppose that an error occurs in the last bit of c, transforming it to $[1110000]$. Before decoding this word, we first check it for errors by multiplying it by the parity check matrix (5.7). The resulting syndrome $s = [001]$ matches the last column of H, indicating that the error has occurred in the last bit of c. So, we correct $[1110000]$ to $[1110001]$ and then decode it by truncating the last three bits. We obtain the data $d = [1110]$.

5.4.8 Lexicographic Parity Check Matrix

If the columns of H are permuted, then the resulting code remains a Hamming code. For example, by permuting the columns of the matrix (5.7), we can get the following parity check matrix:

$$H = \begin{bmatrix} 0\,0\,0\,1\,1\,1\,1 \\ 0\,1\,1\,0\,0\,1\,1 \\ 1\,0\,1\,0\,1\,0\,1 \end{bmatrix}. \tag{5.9}$$

This matrix represents a different $(7, 4)$ Hamming code. Note that each column i corresponds to a binary representation of the integer i, for $i \in \{1, 2, \ldots, 2^{n-k} - 1\}$. A parity check matrix satisfying this property is called the *lexicographic parity check matrix*. The Hamming code corresponding to the matrix (5.9) does not have a generator matrix in the standard form $G = [I_3 A]$.

For a Hamming code with the lexicographic parity check matrix, a very simple procedure for error correction exists. To check a codeword c for errors, we first calculate the syndrome $s = Hc^T$. If s is zero, then no error has occurred. If s is not zero, then we assume a single error has occurred in the ith bit of c corresponding to the decimal representation of the syndrome vector $s = [s_0 s_1 \ldots s_{n-k-1}]$:

$$i = \sum_{0}^{n-k-1} 2^i \times s_i,$$

We correct the error by complementing the ith bit of c.

As we can see, this error correction procedure is simpler than the one described in Sect. 5.4.5, because we do not need to search for the column of H which matches s.

The generator matrix corresponding to the lexicographic parity check matrix (5.9) is given by [14]:

$$G = \begin{bmatrix} 0\,1\,0\,1\,0\,1\,0 \\ 1\,0\,0\,1\,1\,0\,0 \\ 1\,1\,1\,0\,0\,0\,0 \\ 1\,1\,0\,1\,0\,0\,1 \end{bmatrix}. \tag{5.10}$$

The reader can verify that $HG^T = 0$. Since H is not in the form $H = [A^T I_3]$, we cannot construct the above G from H using the standard procedure.

By multiplying the data $d = [d_0 d_1 d_2 d_3]$ by G, we get the codeword

$$c = [c_0 c_1 c_3 c_4 c_5 c_6 c_7] = [p_2 p_1 d_2 p_0 d_1 d_0 d_3],$$

where p_0, p_1, p_2 are check bits defined by $p_0 = d_0 \oplus d_2 \oplus d_3$, $p_1 = d_0 \oplus d_2 \oplus d_3$ and $p_2 = d_1 \oplus d_2 \oplus d_3$. We can see that every data bit is protected by several parity checks. Thus, Hamming codes are a type of overlapping parity codes.

5.4.9 Applications of Hamming Codes

Hamming codes are widely used for error correction in DRAMs. As we mentioned before, DRAMs are susceptible to random bit flips caused by alpha particles from the device packaging, background radiation, etc. A typical fault rate is one single-bit flip per day per gigabyte of DRAM [30].

A memory protected by a Hamming code looks similar to the one in Fig. 5.3 except that more than one parity check bit is generated for each data word. Encoding is performed on complete data words, rather than individual bytes. The check bits are computed by parity generators which are designed according to the generator matrix of the code. For instance, for a (7, 4) Hamming code with the generator matrix (5.10), three parity generators implement the equations $p_0 = d_0 \oplus d_2 \oplus d_3$, $p_1 = d_0 \oplus d_2 \oplus d_3$ and $p_2 = d_1 \oplus d_2 \oplus d_3$. A parity generator is usually realized as a tree of XOR gates.

When a codeword is read back, check bits are recomputed and the syndrome is generated by taking an XOR of the read and recomputed check bits. If the syndrome is zero, no error has occurred. If the syndrome is nonzero, the faulty bit is located by comparing the syndrome to the columns of the parity check matrix H. The position of the matching column in H corresponds to the bit position of the error. The comparison can be implemented either in hardware or in software. The error is corrected by complementing the faulty bit.

If an (n, k) Hamming code with a lexicographic parity check matrix is used, then error correction can be easily implemented in hardware using a decoder and $2n-k-1$ XOR gates [14]. For instance, an error-correcting circuit for (7, 4) Hamming code with the generator matrix (5.10) is shown in Fig. 5.6. The XOR gates in the first level compare read check bits p_i with recomputed check bits p_i^*, for $i \in \{0, 1, 2\}$.

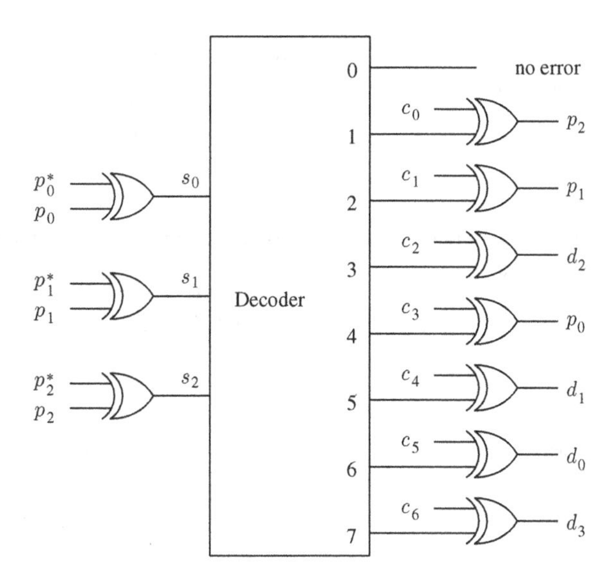

Fig. 5.6 Error-correcting circuit for an (7, 4) Hamming code with a lexicographic parity check matrix

The result is the syndrome $s = [s_0 s_1 s_2]$, which is fed into the 3–7 decoders. For the syndrome $s = i, i \in \{0, 1, \ldots, 7\}$, the ith output of the decoder is high. The XOR gates in the second level complement the ith bit of the codeword, thus correcting the error.

In some applications, the extended Hamming code which allows for not only single-bit error correction, but also double-bit error detection, is used. We describe this code in the next section.

5.4.10 Extended Hamming Codes

The code distance of a Hamming code is 3. If we add a parity check bit to every codeword of a Hamming code, then the code distance increases to 4. The resulting code is called an *extended Hamming code*. It can correct single-bit errors and detect double-bit errors.

The parity check matrix for an extended (n, k) Hamming code can be constructed by first adding a zero column in front of a parity check matrix of an (n, k) Hamming code, and then by attaching a row consisting of all 1's as the $n - k + 1$th row of the resulting matrix. For example, the parity check matrix H for an extended $(1, 1)$ Hamming code can be constructed as:

$$H = \left[\begin{array}{c|c} 0 & 1 \\ \hline 1 & 1 \end{array}\right].$$

The parity check matrix H for an extended (3, 2) Hamming code can be constructed as:

$$H = \left[\begin{array}{c|ccc} 0 & 0 & 1 & 1 \\ 0 & 1 & 0 & 1 \\ \hline 1 & 1 & 1 & 1 \end{array}\right].$$

The parity check matrix H for an extended (7, 4) Hamming code can be constructed as:

$$H = \left[\begin{array}{c|ccccccc} 0 & 0 & 0 & 0 & 1 & 1 & 1 & 1 \\ 0 & 0 & 1 & 1 & 0 & 0 & 1 & 1 \\ 0 & 1 & 0 & 1 & 0 & 1 & 0 & 1 \\ \hline 1 & 1 & 1 & 1 & 1 & 1 & 1 & 1 \end{array}\right].$$

If $c = [c_1, c_2, \ldots, c_n]$ is a codeword of an (n, k) Hamming code, then $c^* = [c_0, c_1, c_2, \ldots, c_n]$ is the corresponding codeword of the extended Hamming code, where $c_0 = \sum_{i=1}^{n} c_i$ is the parity check bit.

5.5 Cyclic Codes

Cyclic codes are a special class of linear codes. Their cyclic nature gives them an additional structure which enables easy encoding and decoding using linear feedback shift registers (LFSRs).

Cyclic codes are usually used in applications where burst errors are dominant. A *burst error* is an error affecting a group of adjacent bits. Such errors are common in data transmission as well as in storage devices, such as disks and tapes. A scratch on a compact disk is an example of a burst error.

5.5.1 Definition

A linear code C is called cyclic if $[c_0 c_1 c_2 \ldots c_{n-2} c_{n-1}] \in C$ implies $[c_{n-1} c_0 c_1 c_2 \ldots c_{n-2}] \in C$. In other words, any end-around shift of a codeword of a cyclic code produces another codeword.

It is convenient to treat codewords of a cyclic code as polynomials rather than vectors. For example, a codeword $c = [c_0 c_1 c_2 \ldots c_{n-1}]$ corresponds to a polynomial

$$c(x) = c_0 \oplus c_1 \cdot x \oplus c_2 \cdot x^2 \oplus \ldots \oplus c_{n-1} \cdot x^{n-1},$$

where $c_i \in \{0, 1\}$ are constants and the addition and the multiplication are in the field $\mathbb{F}_2$, i.e modulo 2. Recall that $x^0 = 1$.

Similarly, a data word $d = [d_0 d_1 d_2 \ldots d_{k-1}]$ corresponds to a polynomial

$$d(x) = d_0 \oplus d_1 \cdot x \oplus d_2 \cdot x^2 \oplus \ldots \oplus d_{k-1} \cdot x^{k-1}.$$

The *degree* of a polynomial is equal to its highest exponent. For example, the polynomial $1 \oplus x^2 \oplus x^3$ has the degree 3.

Before continuing with cyclic codes, we review the basics of polynomial arithmetic, necessary for the understanding of encoding and decoding algorithms.

5.5.2 Polynomial Manipulation

In this section we consider examples of polynomial multiplication and division. All the operations are carried out in the field $\mathbb{F}_2$, i.e. modulo 2.

Example 5.8. Compute $(1 \oplus x \oplus x^3) \cdot (1 \oplus x^2)$.

$$(1 \oplus x \oplus x^3) \cdot (1 \oplus x^2) = 1 \oplus x \oplus x^3 \oplus x^2 \oplus x^3 \oplus x^5 = 1 \oplus x \oplus x^2 \oplus x^5.$$

Note that $x^3 \oplus x^3 = 0$, since the addition is modulo 2.

Example 5.9. Compute $(1 \oplus x^3 \oplus x^5 \oplus x^6)/(1 \oplus x \oplus x^3)$.

$$\begin{array}{r|l}
 & x^3 \oplus x^2 \oplus x \oplus 1 \\ \hline
x^3 \oplus x \oplus 1 & x^6 \oplus x^5 \oplus x^3 \oplus 1 \\
 & x^6 \oplus x^4 \oplus x^3 \\ \hline
 & \quad x^5 \oplus x^4 \oplus 1 \\
 & \quad x^5 \oplus x^3 \oplus x^2 \\ \hline
 & \qquad x^4 \oplus x^3 \oplus x^2 \oplus 1 \\
 & \qquad x^4 \oplus x^2 \oplus x \\ \hline
 & \qquad\quad x^3 \oplus x \oplus 1 \\
 & \qquad\quad x^3 \oplus x \oplus 1 \\ \hline
 & \qquad\qquad 0
\end{array}$$

So, $1 \oplus x \oplus x^3$ divides $1 \oplus x^3 \oplus x^5 \oplus x^6$ without a remainder and the result is $1 \oplus x \oplus x^2 \oplus x^3$.

For the decoding algorithm we also need to perform division modulo $p(x)$, where $p(x)$ is a polynomial. To find $f(x) \bmod p(x)$, we divide the polynomial $f(x)$ by $p(x)$ and take the remainder as a result. If the degree of $f(x)$ is smaller than the degree of $p(x)$, then $f(x) \bmod p(x) = f(x)$.

Example 5.10. Compute $(1 \oplus x^2 \oplus x^5) \bmod (1 \oplus x \oplus x^3)$.

$$
\begin{array}{r|l}
 & x^1 \oplus 1 \\
\hline
x^3 \oplus x \oplus 1 & x^5 \oplus x^2 \oplus 1 \\
 & x^5 \oplus x^3 \oplus x^2 \\
\hline
 & \quad x^3 \oplus 1 \\
 & \quad x^3 \oplus x \oplus 1 \\
\hline
 & \qquad x
\end{array}
$$

So, $(1 \oplus x^2 \oplus x^5) \bmod (1 \oplus x \oplus x^3) = x$.

5.5.3 Generator Polynomial

To encode data in a cyclic code, the polynomial representing the data is multiplied by a polynomial called the *generator polynomial.*

For example, suppose we encode the data $d = [d_0 d_1 d_2 d_3] = [1001]$, using the generator polynomial $g(x) = 1 \oplus x \oplus x^3$. The polynomial representing the data is $d(x) = 1 \oplus x^3$. By multiplying $d(x)$ by $g(x)$, we get the polynomial

$$c(x) = g(x) \cdot d(x) = 1 \oplus x \oplus x^4 \oplus x^6$$

representing the codeword $c = [c_0 c_1 c_2 c_3 c_4 c_5 c_6] = [1100101]$.

If k is the length of the data and $deg(g(x))$ is the degree of the generator polynomial $g(x)$, then the length of the resulting codeword is $n = k + deg(g(x))$, where " + " is the arithmetic addition.

The degree of the generator polynomial determines the error-correcting capabilities of the cyclic code. An (n, k) cyclic code can detect burst errors affecting $n - k$ bits or less.

The choice of the generator polynomial is guided by the following property [21].

Property 5.2. The polynomial $g(x)$ is the generator polynomial for a cyclic code of length n if and only if $g(x)$ divides $1 \oplus x^n$ without a remainder.

So, in order to find a generator polynomial $g(x)$ for an (n, k) cyclic code, we should first factor $1 \oplus x^n$ and then choose a polynomial with the degree $n - k$ among the resulting divisors.

Example 5.11. Find a generator polynomial for a (7, 4) cyclic code. Construct the resulting code.

We are looking for a polynomial of degree $7 - 4 = 3$ which divides $1 \oplus x^7$ without a remainder. The polynomial $1 \oplus x^7$ can be factored as

$$1 \oplus x^7 = (1 \oplus x \oplus x^3)(1 \oplus x^2 \oplus x^3)(1 \oplus x).$$

Table 5.7 Defining table for (7, 4) cyclic code with the generator polynomial $g(x) = 1 \oplus x \oplus x^3$

Data				Codeword						
d_0	d_1	d_2	d_3	c_0	c_1	c_2	c_3	c_4	c_5	c_6
0	0	0	0	0	0	0	0	0	0	0
0	0	0	1	0	0	0	1	1	0	1
0	0	1	0	0	0	1	1	0	1	0
0	0	1	1	0	0	1	0	1	1	1
0	1	0	0	0	1	1	0	1	0	0
0	1	0	1	0	1	1	1	0	0	1
0	1	1	0	0	1	0	1	1	1	0
0	1	1	1	0	1	0	0	0	1	1
1	0	0	0	1	1	0	1	0	0	0
1	0	0	1	1	1	0	0	1	0	1
1	0	1	0	1	1	1	0	0	1	0
1	0	1	1	1	1	1	1	1	1	1
1	1	0	0	1	0	1	1	1	0	0
1	1	0	1	1	0	1	0	0	0	1
1	1	1	0	1	0	0	0	1	1	0
1	1	1	1	1	0	0	1	0	1	1

So we can choose either $g(x) = 1 \oplus x \oplus x^3$ or $g(x) = 1 \oplus x^2 \oplus x^3$. Table 5.7 shows the code generated by $g(x) = 1 \oplus x \oplus x^3$. Since $deg(g(x)) = 3$, this code can detect all burst errors affecting 3 bits or less.

Example 5.12. Find a generator polynomial for a (7, 3) cyclic code.

We are looking for a polynomial of degree $7 - 3 = 4$ which divides $1 \oplus x^7$ without a remainder. Be examining the divisors of the polynomial $1 \oplus x^7$ in the example above, we can see that we can use them to create two polynomials of degree 4: $g(x) = (1 \oplus x \oplus x^3)(1 \oplus x) = 1 \oplus x \oplus x^2 \oplus x^4$ and $g(x) = (1 \oplus x^2 \oplus x^3)(1 \oplus x) = 1 \oplus x^2 \oplus x^3 \oplus x^4$. Since both of them divide $1 \oplus x^7$ without a remainder, we can use either of them to generate a (7, 3) cyclic code.

Let C be an (n, k) cyclic code generated by $g(x)$. Codewords $x^i g(x)$, for $i \in \{0, 1, \ldots, k-1\}$, are a basis for C, since every codeword $c(x)$ is a linear combination of vectors $x^i g(x)$:

$$c(x) = d(x)g(x) = d_0 g(x) \oplus d_1 x g(x) \oplus \ldots d_{k-1} x^{k-1} g(x)$$

Therefore, a matrix with k rows corresponding to vectors $x^i g(x)$, for $i \in \{0, 1, \ldots, k-1\}$, is the generator matrix for C:

$$G = \begin{bmatrix} g(x) \\ xg(x) \\ \ldots \\ x^{k-2}g(x) \\ x^{k-1}g(x) \end{bmatrix} = \begin{bmatrix} g_0\ g_1 \ldots g_{n-k} & 0 & 0 & \ldots & 0 \\ 0\ g_0\ g_1 & \ldots & g_{n-k} & 0 & \ldots & 0 \\ & & \ldots & & & \\ 0\ 0\ 0 & g_0 & g_1 & \ldots\ g_{n-k} & 0 \\ 0\ 0\ 0 & 0 & g_0 & g_1 & \ldots & g_{n-k} \end{bmatrix}.$$

Every row of G is a right cyclic shift of its first row. By examining the structure of G, the reader can see why for cyclic codes the encoding based on the matrix multiplication by G can be replaced by the encoding based on the polynomial multiplication by $g(x)$.

For example, if C is a $(7,4)$ cyclic code with the generator polynomial $g(x) = 1 \oplus x \oplus x^3$, then its generator matrix is in the form:

$$G = \begin{bmatrix} 1\,1\,0\,1\,0\,0\,0 \\ 0\,1\,1\,0\,1\,0\,0 \\ 0\,0\,1\,1\,0\,1\,0 \\ 0\,0\,0\,1\,1\,0\,1 \end{bmatrix}.$$

5.5.4 Parity Check Polynomial

The parity check polynomial $h(x)$ for the cyclic code C is related to the generator polynomial $g(x)$ of C by the equation

$$g(x)h(x) = 1 \oplus x^n.$$

This equation implies that, if the data $d(x)$ are encoded into a codeword $c(x)$ using the generator polynomial $g(x)$, then the product of the parity check polynomial and the codeword should be zero. This is true because every codeword $c(x) \in C$ is a multiple of $g(x)$, so it holds that

$$c(x)h(x) = d(x)g(x)h(x) = d(x)(1 \oplus x^n) = 0 \bmod 1 \oplus x^n.$$

A *parity check matrix* H of a cyclic code contains as its first row the coefficient of $h(x)$, starting from the most significant one. Every following row of H is a right cyclic shift of the first row:

$$H = \begin{bmatrix} h_k & h_{k-1} & \dots & h_0 & 0 & 0 & \dots & 0 \\ 0 & h_k & h_{k-1} & \dots & h_0 & 0 & \dots & 0 \\ & & & \dots & & & & \\ 0 & \dots & 0 & h_k & h_{k-1} & \dots & h_0 & 0 \\ 0 & \dots & 0 & 0 & h_k & h_{k-1} & \dots & h_0 \end{bmatrix}.$$

Example 5.13. Let C be an $(7,4)$ cyclic code with the generator polynomial $g(x) = 1 \oplus x \oplus x^3$. Compute its parity check polynomial.

Since $1 \oplus x^7$ is factored as:

$$1 \oplus x^7 = (1 \oplus x \oplus x^3)(1 \oplus x^2 \oplus x^3)(1 \oplus x) = g(x)(1 \oplus x^2 \oplus x^3)(1 \oplus x)$$

we can conclude that $h(x) = (1 \oplus x^2 \oplus x^3)(1 \oplus x) = 1 \oplus x \oplus x^2 \oplus x^4$. The corresponding parity check matrix is given by

$$H = \begin{bmatrix} 1\,0\,1\,1\,1\,0\,0 \\ 0\,1\,0\,1\,1\,1\,0 \\ 0\,0\,1\,0\,1\,1\,1 \end{bmatrix}. \quad (5.11)$$

5.5.5 Syndrome Polynomial

From the discussion in the previous section, we can conclude that the *syndrome polynomial* can be defined as

$$s(x) = h(x)c(x) \bmod 1 \oplus x^n.$$

Furthermore, since $g(x)h(x) = 1 \oplus x^n$, we can compute $s(x)$ by dividing the $c(x)$ by $g(x)$.

If $s(x) = 0$ is zero, no error has occurred. Otherwise, if the coefficients $[s_0 s_1 \ldots s_{n-k-1}]$ match one of the columns of parity check matrix H, then a single-bit error has occurred. The bit position of the error corresponds to the position of the matching column in H, so the error can be corrected. If the syndrome is not zero and it is not equal to any of the columns of H, then a multiple-bit error has occurred which cannot be corrected.

5.5.6 Implementation of Encoding and Decoding

In this section, we show how encoding and decoding algorithms for cyclic codes can be efficiently implemented using LFSRs.

5.5.6.1 Encoding by Polynomial Multiplication

The multiplication of the data polynomial $d(x) = d_0 \oplus d_1 x \oplus d_2 x^2 \oplus \ldots \oplus d_{k-1} x^{k-1}$ by the generator polynomial $g(x) = g_0 \oplus g_1 x \oplus g_2 x^2 \oplus \ldots \oplus g_r x^r$, where $r = n - k$, can be implemented using an LFSR, as shown in Fig. 5.7.

In the diagram in Fig. 5.7, the square boxes are register cells, each capable of storing one bit of information. The circles labeled by "$\oplus$" are modulo 2 adders. The triangles represent the weights corresponding to the coefficients g_i of the generator polynomial $g(x)$, for $i \in \{0, 1, \ldots, r\}$. Each coefficient g_i is either 0, meaning "no connection", or 1, meaning "connection". The clock signal is not shown at the diagram. At each clock cycle, the register cells are loaded with the values at their inputs [9].

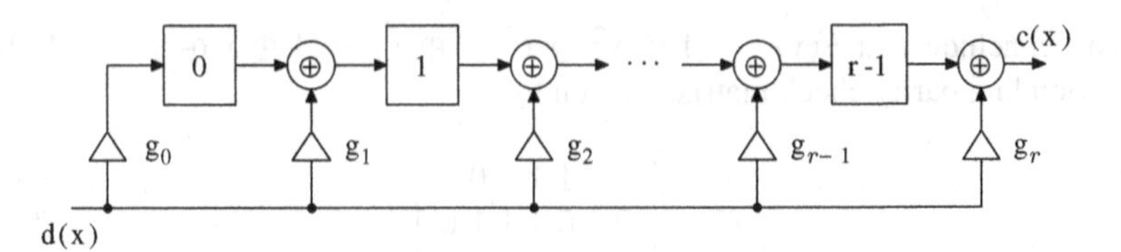

Fig. 5.7 Logic diagram of an LFSR multiplying the data polynomial $d(x)$ by the generator polynomial $g(x)$. The result is the codeword polynomial $c(x)$

The multiplication is carried out as follows [21]. Initially, all register cells are set to 0. The coefficients of the data and codeword polynomial appear on the input and output lines *high-order coefficient first*. As the first data coefficient d_{k-1} appears at the input, the output becomes $d_{k-1}g_r$, which is the correct coefficient c_{n-1} of the codeword polynomial. After the shift, the register cells are loaded with $d_{k-1}g_0, d_{k-1}g_1, \ldots, d_{k-1}g_{r-1}$, the input becomes d_{k-2} and the output becomes $d_{k-1}g_{r-1} \oplus d_{k-2}g_r$, which is the correct coefficient c_{n-2}. At the next clock cycle, the content of the register cells becomes $d_{k-2}g_0, d_{k-1}g_0 \oplus d_{k-2}g_1, d_{k-1}g_1 \oplus d_{k-2}g_2, \ldots, d_{k-1}g_{r-2} \oplus d_{k-2}g_{r-1}$, the input becomes d_{k-3}, and the output becomes $d_{k-1}g_{r-2} \oplus d_{k-2}g_{r-1} \oplus d_{k-3}g_r$, which is the correct coefficient c_{n-3}. The multiplication process continues in a similar way.

The reader is advised to write out the contents of the register cells at each step and to compare them with the sequence of intermediate results obtained during the ordinary hand calculation of the polynomial multiplication.

As an example, consider the LFSR shown in Fig. 5.8. It implements the multiplication by the generator polynomial $g(x) = 1 \oplus x \oplus x^3$. Modulo 2 adders are realized using XOR gates.

Let $s_i \in \{0, 1\}, i \in \{0, 1, 2\}$, represent the value of the register cell i at the current time step. The vector of values of register cells is called a *state* of the LFSR. Let $s_i^+ \in \{0, 1\}$ represent the value of the register cell i at the next time step. Then, from the circuit in Fig. 5.10, we can derive the following next-state equations:

$$\begin{aligned} s_0^+ &= d(x) \\ s_1^+ &= s_0 \\ s_2^+ &= s_1 \oplus d(x) \\ c(x) &= s_2 \oplus d(x). \end{aligned}$$

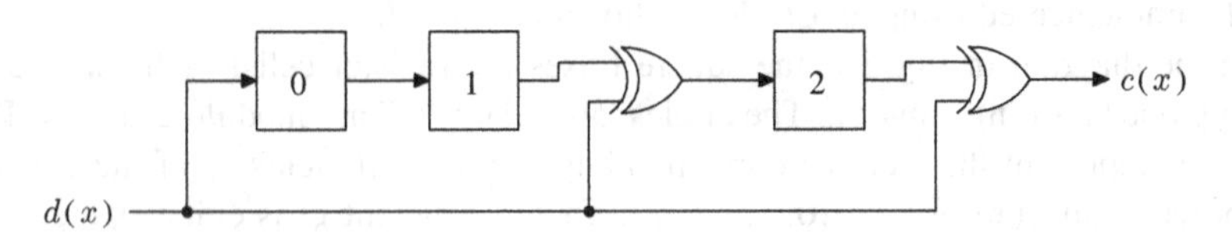

Fig. 5.8 An LFSR implementing the multiplication by the generator polynomial $g(x) = 1 \oplus x^2 \oplus x^3$

Table 5.8 Sequence of states of the LFSR in Fig. 5.8 for the input $d = [0101]$

Clock Cycle	Input $d(x)$	Next LFSR state s_0^+	s_1^+	s_2^+	Output $c(x)$	
		0	0	0		
1	1	1	0	1	1	x^6
2	0	0	1	0	1	x^5
3	1	1	0	0	1	x^4
4	0	0	1	0	0	
5	0	0	0	1	0	
6	0	0	0	0	1	x
7	0	0	0	0	0	

Table 5.8 illustrates the encoding process for the data $d = [d_0 d_1 d_2 d_3] = [0101]$. The resulting codeword is $c = [c_0 c_1 c_2 c_4 c_5 c_6] = [0100111]$. It is easy to verify that by multiplying $d(x)$ by $g(x)$:

$$d(x) \cdot g(x) = (x \oplus x^3)(1 \oplus x^2 \oplus x^3) = x \oplus x^4 \oplus x^5 \oplus x^6 = c(x).$$

5.5.6.2 Decoding by Polynomial Division

The division of the codeword polynomial $c(x) = c_0 \oplus c_1 x \oplus c_2 x^2 \oplus \ldots \oplus c_{n-1} x^{n-1}$ by the generator polynomial $g(x) = g_0 \oplus g_1 x \oplus g_2 x^2 \oplus \ldots \oplus g_r x^r$, where $r = n - k$, can be implemented by an LFSR, as shown in Fig 5.9. As previously, the weights $g_i, i \in \{0, 1, \ldots, r-1\}$, correspond to the coefficients of the generator polynomial $g(x)$. The weight 0 means "no connection" and the weight 1 means "connection". An exception is g_r which is always connected.

The division is carried out as follows [21]. Initially, all register cells are set to 0. The coefficients of the codeword and data polynomials appear on the input and output lines, high-order coefficient first. It takes $r - 1$ clock cycles for the first input coefficient to reach the last register cell of the LFSR. Therefore, the output is 0 for the first $r - 1$ clock cycles. Afterwards, $c_{n-1} g_r$ appears at the output. It corresponds to the coefficient d_{k-1} of the data polynomial. For each nonzero quotient coefficient $d_i, i \in \{0, 1, \ldots, k-1\}$, the polynomial $d_i g(x)$ is subtracted from the dividend. The feedback connections accomplish this subtraction. Note that the subtraction modulo 2 is equivalent to the addition modulo 2 in $\mathbb{F}_2$. After a total of n clock cycles, all coefficients of the data polynomial are computed. The register cells contain the remainder $[s_0 s_1 \ldots s_{r-1}]$, which corresponds to the syndrome.

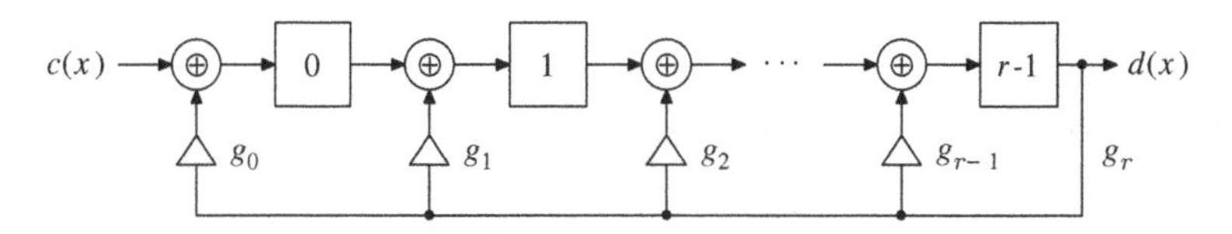

Fig. 5.9 Logic diagram of an LFSR dividing the codeword polynomial $c(x)$ by the generator polynomial $g(x)$. The result is the data polynomial $d(x)$

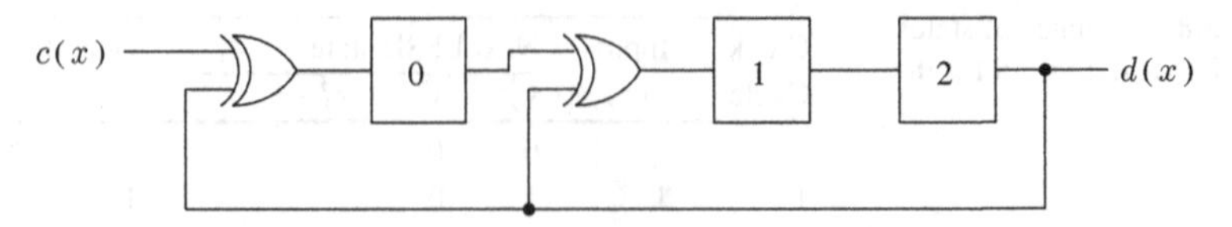

Fig. 5.10 An LFSR implementing the division by the generator polynomial $g(x) = 1 \oplus x \oplus x^3$

The reader is advised to write out the contents of the register cells at each step and to compare them with the sequence of intermediate results obtained during the ordinary hand calculation of the polynomial division.

As an example, consider the LFSR shown in Fig. 5.10. It implements the division by the generator polynomial $g(x) = 1 \oplus x \oplus x^3$. The next-state equations for this LFSR are given by

$$\begin{aligned} s_0^+ &= s_2 \oplus c(x) \\ s_1^+ &= s_0 \oplus s_2 \\ s_2^+ &= s_1 \\ d(x) &= s_2. \end{aligned}$$

Suppose the word to be decoded is $c = [c_0c_1c_2c_4c_5c_6] = [1010001]$. Table 5.9 illustrates the decoding process. The coefficients of $c(x)$ are fed into the LFSR starting from the high-order coefficient c_6. The first bit of the quotient, corresponding to the high-order coefficient of the data, d_3, appears at the output line at the 4th clock cycle. The division is completed after seven cycles. The state of the LFSR after seven cycles is $s = [s_0s_1s_2] = [000]$. Since the syndrome is 0, there is no error. So, $c(x)$ is a proper codeword and $d = [1101]$ is valid data. We can verify that by dividing $c(x)$ by $g(x)$:

$$c(x)/g(x) = (1 \oplus x^2 \oplus x^6)/(1 \oplus x \oplus x^3) = 1 \oplus x \oplus x^3 = d(x).$$

Next, suppose that a single-bit error has occurred in the 4th bit of codeword $c = [1010001]$ and a word $c^* = [1011001]$ is received instead. Table 5.10 illustrates the decoding process. As we can see, after the division is completed, the register cells contain the syndrome $s = [110]$, which matches the 4th column of the parity check matrix H listed as (5.11).

We can verify the result by dividing $c^*(x)$ by $g(x)$:

$$c^*(x)/g(x) = (1 \oplus x^2 \oplus x^3 \oplus x^6)/(1 \oplus x \oplus x^3) = (x \oplus x^3) \oplus (1 \oplus x)/(1 \oplus x \oplus x^3).$$

5.5.7 Separable Cyclic Codes

The cyclic codes which we have studied so far were not separable. It is possible to construct a separable cyclic code using the following technique [21].

Table 5.9 Sequence of states of the LFSR in Fig. 5.10 for the input codeword [1010001]

Clock Cycle	Input $c(x)$	Next LFSR state s_0^+	s_1^+	s_2^+	Output $d(x)$	
		0	0	0		
1	1	1	0	0	0	
2	0	0	1	0	0	
3	0	0	0	1	0	
4	0	1	1	0	1	x^3
5	1	1	1	1	0	
6	0	1	0	1	1	x
7	1	0	0	0	1	1

Table 5.10 Sequence of states of the LFSR in Fig. 5.10 for the input word [1011001]

Clock Cycle	Input $c(x)$	Next LFSR state s_0^+	s_1^+	s_2^+	Output $d(x)$	
0		0	0	0		
1	1	1	0	0	0	
2	0	0	1	0	0	
3	0	0	0	1	0	
4	1	0	1	0	1	x^3
5	1	1	0	1	0	
6	0	1	0	0	1	x
7	1	1	1	0	0	

First, we take the data $d = [d_0 d_1 \ldots d_{k-1}]$ to be encoded and shift it right by $n-k$ positions. The result is the following n-bit word:

$$[00\ldots 0 d_0 d_1 \ldots d_{k-1}]$$

Shifting the vector d right by $n-k$ positions corresponds to multiplying the data polynomial $d(x)$ by the term x^{n-k}. So, the above n-bit word corresponds to the following polynomial:

$$d(x)x^{n-k} = d_0 x^{n-k} \oplus d_1 x^{n-k+1} \oplus \ldots \oplus d_{k-1} x^{n-1}$$

The polynomial $d(x)x^{n-k}$ can be decomposed as:

$$d(x)x^{n-k} = q(x)g(x) \oplus r(x)$$

where $q(x)$ is the quotient and $r(x)$ is the remainder of division. The remainder $r(x)$ has a degree less than $n-k$, i.e. it is of type

$$[r_0 r_1 \ldots r_{n-k-1} 00 \ldots 0]$$

By moving $r(x)$ from the right-hand side of the equation to the left-hand side of the equation we get:

$$d(x)x^{n-k} \oplus r(x) = q(x)g(x)$$

Recall that subtraction modulo 2 is equivalent to addition modulo 2 in $\mathbb{F}_2$. Since the left-hand side of this equation is a multiple of $g(x)$, it is a codeword. This codeword has the form

$$[r_0 r_1 \dots r_{n-k-1} d_0 d_1 \dots d_k]$$

So, we have obtained a codeword in which the data are separated from the check bits, which was our goal.

Let us illustrate the encoding technique explained above in the example of the (7,4) cyclic code with the generator polynomial $g(x) = 1 \oplus x \oplus x^3$. Suppose that the data to be encoded is $d(x) = x \oplus x^3$, i.e. $d = [d_0 d_1 d_2 d_3] = [0101]$.

First, we compute $x^{n-k}d(x)$. Since $n - k = 3$, we get $x^3(x \oplus x^3) = x^4 \oplus x^6$. by decomposing $x^4 \oplus x^6$, we get:

$$x^4 \oplus x^6 = (1 \oplus x^3)(1 \oplus x \oplus x^3) \oplus (1 \oplus x)$$

So, the remainder $r(x)$ is $r(x) = 1 \oplus x$ and the resulting codeword is

$$c(x) = d(x)x^{n-k} \oplus r(x) = 1 \oplus x \oplus x^4 \oplus x^6$$

i.e. $c = [c_0 c_1 c_2 c_3 c_4 c_5 c_6] = [1100101]$. The data are contained in the last four bits of the codeword.

Example 5.14. Construct the generator and parity check matrices for a separable (7,4) cyclic code with the generator polynomial $g(x) = 1 \oplus x \oplus x^3$.

From Example 5.13, we know that a (7,4) cyclic code with the generator polynomial $g(x) = 1 \oplus x \oplus x^3$ has the parity check matrix given by (5.11). Since the columns of H represent all possible nonzero binary vectors of length 3, we can conclude that this code is a Hamming code.

In Sect. 5.4.7, we showed that a separable Hamming code can be obtained from a nonseparable one by permuting the columns of H to bring it into the form $H = [A^T I_{n-k}]$. We can apply the same technique to the matrix (5.7). One of the possible solutions is:

$$H = \begin{bmatrix} 1\,0\,1\,1\,1\,0\,0 \\ 1\,1\,1\,0\,0\,1\,0 \\ 0\,1\,1\,1\,0\,0\,1 \end{bmatrix}.$$

The corresponding generator matrix $G = [I_k A]$ is:

$$G = \begin{bmatrix} 1\,0\,0\,0\,1\,1\,0 \\ 0\,1\,0\,0\,0\,1\,1 \\ 0\,0\,1\,0\,1\,1\,1 \\ 0\,0\,0\,1\,1\,0\,1 \end{bmatrix}.$$

Since the encoding of a separable cyclic code involves polynomial division, it can be implemented using an LFSR similar to the one used for decoding. The input of the LFSR is the n-bit vector $[00\ldots 0d_0d_1\ldots d_{k-1}]$ representing the coefficients of the polynomial $x^{n-k}d(x)$. After n clock cycles, when the last bit of $d(x)$ has been fed in, the LFSR contains the remainder of division of the input polynomial $x^{n-k}d(x)$ by the generator polynomial $g(x)$. By adding this remainder to $x^{n-k}d(x)$, we obtain the resulting codeword.

5.5.8 Cyclic Redundancy Check Codes

Cyclic redundancy check (CRC) codes are separable cyclic codes with specific generator polynomials, chosen to provide high error-detection capability for data transmission and storage [20]. Common generator polynomials for CRC are:

- CRC-12: $1 \oplus x \oplus x^2 \oplus x^3 \oplus x^{11} \oplus x^{12}$
- CRC-16: $1 \oplus x^2 \oplus x^{15} \oplus x^{16}$
- CRC-CCITT: $1 \oplus x^5 \oplus x^{12} \oplus x^{16}$
- CRC-32: $1 \oplus x \oplus x^2 \oplus x^4 \oplus x^7 \oplus x^8 \oplus x^{10} \oplus x^{11} \oplus x^{12} \oplus x^{16} \oplus x^{22} \oplus x^{23} \oplus x^{26} \oplus x^{32}$
- CRC-64-ISO: $1 \oplus x \oplus x^3 \oplus x^4 \oplus x^{64}$
- CRC-64-ECMA-182: $1 \oplus x \oplus x^4 \oplus x^7 \oplus x^9 \oplus x^{10} \oplus x^{12} \oplus x^{13} \oplus x^{17} \oplus x^{19} \oplus x^{21} \oplus x^{22} \oplus x^{23} \oplus x^{24} \oplus x^{27} \oplus x^{29} \oplus x^{31} \oplus x^{32} \oplus x^{33} \oplus x^{35} \oplus x^{37} \oplus x^{38} \oplus x^{39} \oplus x^{40} \oplus x^{45} \oplus x^{46} \oplus x^{47} \oplus x^{52} \oplus x^{54} \oplus x^{55} \oplus x^{57} \oplus x^{62} \oplus x^{64}$

The CRC-12 is used for transmission of streams of 6-bit characters in telecom systems. CRC-16 and CRC-CCITT are used for 8-bit transmission streams in modems and network protocols in the USA and Europe, respectively, and give adequate protection for most applications. An attractive feature of CRC-16 and CRC-CCITT is the small number of nonzero terms in their polynomials. This is an advantage, because the LFSR required to implement encoding and decoding is simpler for generator polynomials with a smaller number of terms. Applications that need extra protection, such as Department of Defense applications, use 32- or 64-bit CRCs.

The encoding and decoding is done either in software or in hardware, using the procedure from Sect. 5.5.7. To perform an encoding, the data polynomial is first shifted right by $deg(g(x))$ bit positions, and then divided by the generator polynomial. The coefficients of the remainder form the check bits of the CRC codeword. The number of check bits is equal to the degree of the generator polynomial. So, a CRC detects all burst errors of length less or equal to $deg(g(x))$. A CRC also detects many errors which are larger than $deg(g(x))$. For example, apart from detecting all burst errors of length 16 or less, CRC-16 and CRC-CCITT are also capable of detecting 99.997 % of burst errors of length 17 and 99.9985 burst errors of length 18 [33].

Applications requiring high data transmission rates usually use CRC computation methods which process 2^k, $k > 1$, rather than one bit per clock cycle [8, 27].

5.5.9 Reed-Solomon Codes

Reed–Solomon codes [23] are a class of separable cyclic codes used to correct errors in a wide range of applications, including storage devices (tapes, compact disks, DVDs, barcodes, and RAID 6), wireless communication (cellular telephones, microwave links), deep space and satellite communication, digital television (DVB, ATSC), and high-speed modems (ADSL, xDSL).

Formally, Reed–Solomon codes are not binary codes. The theory behind Reed–Solomon codes is a finite field $\mathbb{F}_2^m$ of degree m over $\{0, 1\}$. The elements of such a field are m-tuples of 0 and 1. For example, for $m = 3$, $\mathbb{F}_2^3 = \{000, 001, 010, 011, 100, 10, 111\}$. Therefore, in a Reed–Solomon code, symbols of the code are m-bit vectors. Usually $m = 8$, i.e. a byte.

An encoder for an (n, k) Reed–Solomon code takes k data symbols of m bits, computes the $n - k$ parity symbols of m bits, and appends the parity symbols to the k data symbols to get a codeword containing n symbols of m bits each. The encoding procedure is similar to the one described in Sect. 5.5.7, except that the computation is carried out on m-bit vectors rather than individual bits. The maximum codeword length is related to m as $n = 2^m - 1$. For example, a Reed–Solomon code operating on 8-bit symbols may have $n = 2^8 - 1 = 255$ symbols per codeword. A Reed–Solomon code can correct up to $\lfloor n - k \rfloor / 2$ symbols that contain errors.

For example, a popular Reed–Solomon code is RS(255,223) where symbols are 8 bits long. Each codeword contains 255 bytes, of which 223 bytes are data and 32 bytes are check symbols. So, $n = 255$, $k = 223$ and therefore this code can correct up to 16 bytes containing errors. Each of these 16 bytes can have multiple-bit errors.

The decoding of Reed–Solomon codes is performed using an algorithm designed by Berlekamp [4]. The popularity of Reed–Solomon codes is due to a large extent to the efficiency of this algorithm. Berlekamp's algorithm was used by Voyager II for transmitting pictures of outer space back to Earth [19]. It is also a basis for decoding compact discs in players. Many additional improvements were made over the years to make Reed–Solomon code practical. Compact discs, for example, use a modified version of Reed–Solomon code called *cross-interleaved* Reed–Solomon code [11].

5.6 Unordered Codes

In this section, we consider another interesting family of codes, called *unordered codes*. Unordered codes are designed to detect a special kind of errors called *unidirectional errors*. A unidirectional error changes either 0s of the word to 1 s, or 1 s of the word to 0s, but not both. For example, a unidirectional error can change a word [1011000] to the word [0001000]. Any single-bit error is unidirectional.

Originally, unordered codes were invented as transmission codes for error detection in asymmetric channels, in which errors affecting bits in only one direction are more likely to occur [3]. Later, they found a variety of other applications, including barcode design [32], online error detection in logic circuits [13], generation of test patterns [29], and construction of frequency hopping lists for use in GSM networks [26].

The name *unordered* codes originate from the following. We say that two binary n-bit vectors $x = [x_0 x_1 \dots x_{n-1}]$ and $y = [y_0 y_1 \dots y_{n-1}]$ are *ordered* if either $x_i \leq y_i$ for all $i \in \{0, 1, \dots, n-1\}$, or $x_i \geq y_i$ for all i. For example if $x = [0101]$ and $y = [0000]$ then x and y are ordered, namely $x \geq y$. An unordered code is a code satisfying the property that any two of its codewords are unordered.

The ability of unordered codes to detect all unidirectional errors follows directly from the above property. A unidirectional error always changes a word x to a word y which is either smaller or greater than x. A unidirectional error cannot change a word x to a word which is not ordered with x. Therefore, if every pair of codewords in a code is unordered, then no unidirectional error can change a codeword into another codeword. Thus, an unordered code detects any unidirectional error.

In this section, we describe two popular types of unordered codes: m-of-n codes and Berger codes.

5.6.1 *M-of-N Codes*

An *m-of-n code* (also called *constant-weight code*) consists of all n-bit words with exactly m 1's. Any d-bit unidirectional error causes the affected codeword to have either $m + d$ of $m - d$ 1's. Therefore, such an error is detected.

A popular class of m-of-n codes is 2-of-5 codes. Such codes are used for representing decimal digits using five bits. Each bit is assigned a weight. The sum of weights corresponds to the represented decimal digit. The weights are assigned, so that the representation of most digits is unique. It is not possible to represent 0 as a sum of two weights, therefore 0 is treated separately. Table 5.11 shows examples of weight assignments used in two barcode systems: Code 25 Industrial, used in

Table 5.11 Examples of weight assignments for 2-of-5 code

Digit	2-of-5 representation of the digit	
	Code 25 industrial 0-1-2-3-6	POSTNET 7-4-2-1-0
0	01100	11000
1	11000	00011
2	10100	00101
3	10010	00110
4	01010	01001
5	00110	01010
6	10001	01100
7	01001	10001
8	00101	10010
9	00011	10100

photofinishing and warehouse sorting industries [31], and Postal Numeric Encoding Technique (POSTNET), used by the United States Post Office [32].

We can see from the table that, in the former case, the weights are assigned as 0-1-2-3-6. For example, the digit 4 is represented as [01010]. In the latter case, the weights are assigned as 7-4-2-1-0. In this case, the digit 4 is represented as [01001].

Another popular class of m-of-n codes is 1-of-n codes (also called *one-hot encoding*). They encode $\log_2 n$ data bits in an n-bit codeword. For example, 1-of-2 code encodes the data [0] and [1] into codewords [01] and [10], respectively. 1-of-4 code encodes the data [00], [01], [10], and [11] into codewords [0001], [0010], [0100], and [1000], respectively. Applications of 1-of-n codes include state assignment for finite-state machine synthesis [6], dual-rail encoding in asynchronous circuits [16], and pulse-position modulation in optical communication systems [22].

An easy way to construct an m-of-n code is to take the original k bits of data $d = [d_0 d_1 \ldots d_{k-1}]$, to complement them as $[\overline{d}_0 \overline{d}_1 \ldots \overline{d}_{k-1}]$ and then to create the $2k$-bit codeword by appending the complemented bits to the original data:

$$c = [c_0 c_1 \ldots c_{2k}] = [d_0 d_1 \ldots d_{k-1} \overline{d}_0 \overline{d}_1 \ldots \overline{d}_{k-1}].$$

For example, the 3-of-6 code is shown in Table 5.12. The reader can easily verify that all codewords have exactly three 1's.

An advantage of $2k$-of-k codes is their separability, which simplifies the encoding and decoding procedures. A disadvantage of $2k$-of-k codes is their low information rate, which is only 1/2.

5.6.2 Berger Codes

Check bits in a Berger code represent the number of 1's in the data word [3]. A codeword of a Berger code of length n consists of k data bits followed by m check bits, where

$$m = \lceil \log_2(k+1) \rceil.$$

Table 5.12 Defining table for the 3 -of- 6 code

Data			Codeword					
d_0	d_1	d_2	c_0	c_1	c_2	c_3	c_4	c_5
0	0	0	0	0	0	1	1	1
0	0	1	0	0	1	1	1	0
0	1	0	0	1	0	1	0	1
0	1	1	0	1	1	1	0	0
1	0	0	1	0	0	0	1	1
1	0	1	1	0	1	0	1	0
1	1	0	1	1	0	0	0	1
1	1	1	1	1	1	0	0	0

Table 5.13 Defining table for the Berger code for 3-bit data

Data			Codeword				
d_0	d_1	d_2	c_0	c_1	c_2	c_3	c_4
0	0	0	0	0	0	1	1
0	0	1	0	0	1	1	0
0	1	0	0	1	0	1	0
0	1	1	0	1	1	0	1
1	0	0	1	0	0	1	0
1	0	1	1	0	1	0	1
1	1	0	1	1	0	0	1
1	1	1	1	1	1	0	0

Table 5.14 Information rate of Berger codes for different data lengths

Number of data bits	Number or check bits	Information rate
4	3	0.57
8	4	0.67
16	5	0.76
32	6	0.84
64	7	0.90
128	8	0.94

The Berger codes with $k = 2^m - 1$ are called *maximal length* Berger codes.

Check bits are computed by complementing the m-bit binary representation of the number of 1's in the data word. For example, if the data are 7 bits, then the number of check bits is $m = 3$. To compute check bits for the data $d = [0110101]$, we count the number of 1's in the data, which is 4, then write the binary representation for 4 in 3 bits, which is 010, and finally complement 010, which is 101. The resulting 10-bit codeword is $c = [0110101101]$.

Table 5.13 shows the Berger code for 3-bit data.

The information rate of the Berger code for k-bit data is $k/(\lceil k + \lceil \log_2(k+1)\rceil)$, which is the highest among separable unordered codes [15]. Table 5.14 shows information rates of Berger codes for different data lengths.

Next, we show how a Berger code can be used for *on-line* (or *concurrent*) error detection in logic circuits.

Consider the full-adder circuit shown in Fig. 5.11. The truth tables of Boolean functions for the sum s and carry out c_{out} outputs are shown in Table 5.15.

We treat the 2-bit output vector $[s\ c_{out}]$ as the data to be encoded. Since $k = 2$, the number of check bits to attach is $m = \lceil \log_2 3 \rceil = 2$. For each of the four possible output vectors [00], [01], [10], [11], we count the number of 1's in the vector, write the resulting number into the 2-bit binary representation, and then complement it. The resulting check bits $[b_0 b_1]$ are shown in the last two columns of Table 5.15.

Now we can interpret the last two columns of Table 5.15 as truth tables of 3-variable Boolean functions which predict the values of check bits. We can write the

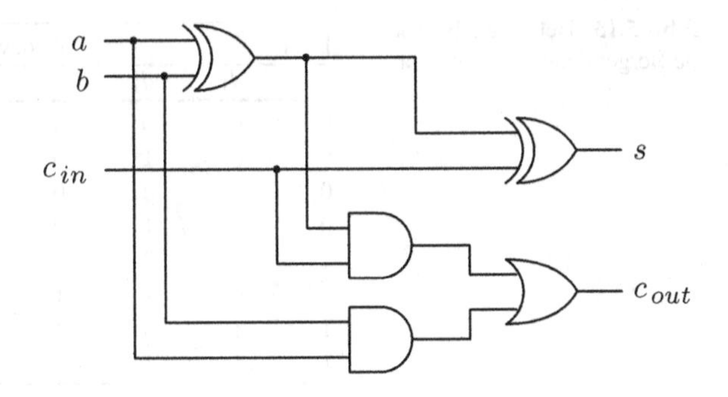

Fig. 5.11 Logic circuit implementing a full-adder

Table 5.15 The truth tables of Boolean functions for the sum s and carry out c_{out} outputs of a full-adder and Boolean functions predicting the check bits b_0 and b_1

Inputs			Outputs		Check bits	
a	b	c_{in}	s	c_{out}	b_0	b_1
0	0	0	0	0	1	0
0	0	1	1	0	0	1
0	1	0	1	0	0	1
0	1	1	0	1	0	1
1	0	0	1	0	0	1
1	0	1	0	1	0	1
1	1	0	0	1	0	1
1	1	1	1	1	0	0

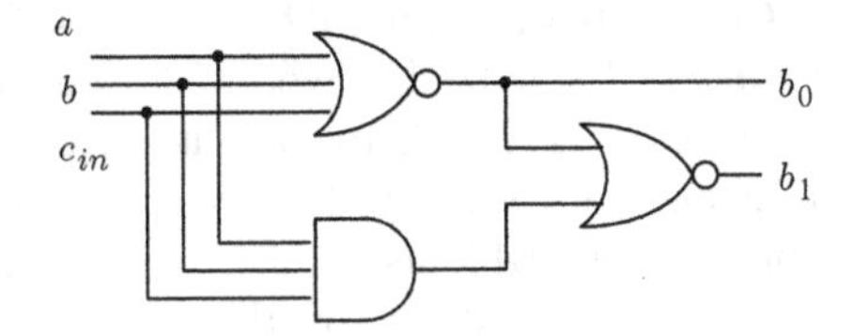

Fig. 5.12 Logic circuit implementing Boolean functions predicting the check bits b_0 and b_1

following Boolean expressions for these functions:

$$b_0(a, b, c_{\text{in}}) = \overline{a}\,\overline{b}\,\overline{c}_{\text{in}}$$

$$b_1(a, b, c_{\text{in}}) = \overline{abc_{\text{in}} + \overline{a}\,\overline{b}\,\overline{c}_{\text{in}}}.$$

The logic circuit which implements these functions is shown in Fig. 5.12.

What remains is to design a checker which re-computes the check bits from the values of the outputs s and c_{out} and compares them with the predicted check bits. The truth tables of Boolean functions for the recomputed check bits are shown in Table 5.16. We can write the following Boolean expressions for these functions:

$$b_0^*(s, c_{\text{out}}) = \overline{s}\,\overline{c}_{\text{out}}$$

$$b_1^*(s, c_{\text{out}}) = s \oplus c_{\text{out}}$$

Table 5.16 The truth tables of Boolean functions for the recomputed check bits b_0^* and b_1^*

s	c_{out}	b_0^*	b_1^*
0	0	1	0
0	1	0	1
1	0	0	1
1	1	0	0

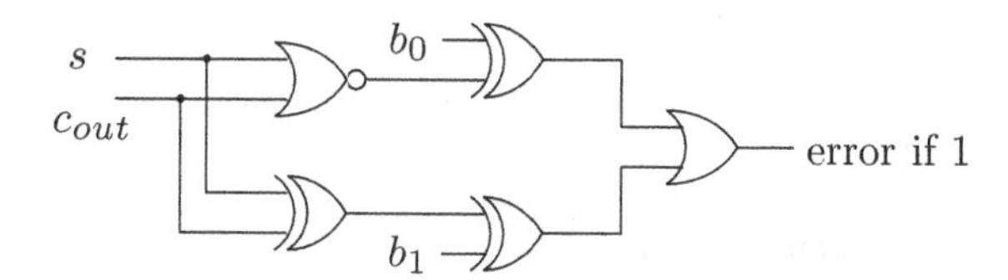

Fig. 5.13 Logic circuit of a checker which re-computes check bits and compares them to the predicted check bits b_0 and b_1

The logic circuit which implements these functions and compares them to the predicted check bits is shown in Fig. 5.13. The output of either of the XOR gates in the second level of logic is 1 if its inputs disagree. Therefore, the output of the OR gate is 1 if at least one of the recomputed check bits disagrees with the predicted check bit.

Now we can combine all blocks into a diagram, as shown in Fig. 5.14. The block labeled "Circuit" corresponds to the full adder in Fig. 5.11. The block labeled "Predictor" represents the circuit in Fig. 5.12. The block labeled "Checker" represents the circuits in Fig. 5.13. The resulting circuit is protected against any faults which cause a unidirectional error in the output vector $[s\ c_{out} b_0 b_1]$. Note that the checker itself is not protected against faults. There are techniques for designing self-checking checkers [15, 18] which can be applied to protect the checker as well.

It is possible to increase the probability that faults in a circuit cause unidirectional errors on the outputs. This can be done by re-synthesizing the original circuit and the check bit predictor, so that they consist of AND and OR gates only, and that inverters appear on the primary inputs only. Logic circuits synthesized in such a way are called *internally monotonic*. If a circuit is internally monotonic, then all paths connecting any internal line to the circuit outputs have the same number of NOT gates, namely zero. Therefore, a single stuck-at fault at any internal line may cause only unidirectional errors on the outputs [13]. Only single stuck-at faults at primary inputs, which contain inverters, may potentially result in bi-directional errors on the outputs. Note, however, that re-synthesizing a circuit into an internally monotonic circuit may increase its area considerably. Therefore, such a technique is not commonly used in practice.

The technique for online error-detection in logic circuits illustrated in Fig. 5.14 can be used not only with a Berger code, but also with any separable error-detecting code, e.g. parity or $2k$-of-k. Predictors and checkers can be constructed according to the rules of the code.

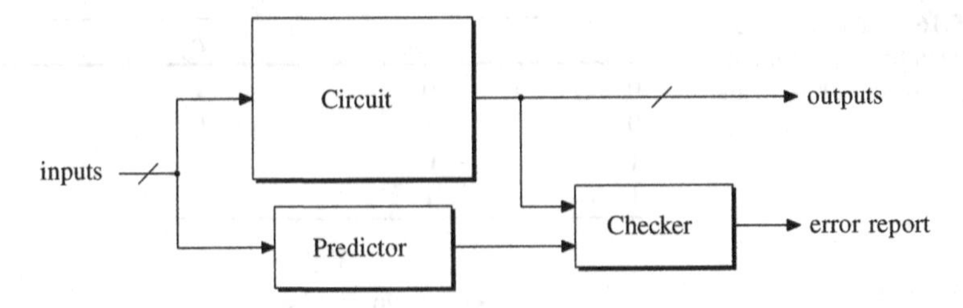

Fig. 5.14 On-line error detection in logic circuits based on separable codes

5.7 Arithmetic Codes

Arithmetic codes are usually used for detecting errors in arithmetic operations, such as addition or multiplication. The operands, say a and b, are encoded before the operation is performed. The operation is carried out on the resulting codewords $A(a)$ and $A(b)$. After the operation, the codeword $A(a) * A(b)$ representing the result of the operation "$*$" is decoded and checked for errors.

An arithmetic code relies on the property of *invariance* with respect to the operation "$*$":

$$A(a * b) = A(a) * A(b)$$

Invariance guarantees that the operation "$*$" on codewords $A(a)$ and $A(b)$ gives us the same result as $A(a * b)$. Therefore, if no error has occurred, the decoding of $A(a * b)$ gives us $a * b$, the result of the operation "$*$" on a and b.

Arithmetic codes differ from other codes in that they use arithmetic distance rather than Hamming distance as a measure of difference between codewords. The arithmetic distance matches more closely the types of errors that may occur during arithmetic operations. In an adder, a single error in one bit may cause errors in several bits due to carry propagation. An error in a carry may also affect multiple bits in the result. Such errors are still counted as single errors under the definition of arithmetic distance.

Let

$$N = \sum_{i=0}^{n-1} x_i \times 2^i$$

be the integer corresponding to the decimal representation of the n-bit binary word $[x_0 x_1 \ldots x_{n-1}]$ (please note that the least significant bit is on the left). The *arithmetic weight* of N in base 2, denoted by $wt(N)$, is defined as the minimum number of nonzero terms in an expression of N in the form

$$N = \sum_{i=0}^{n} a_i \times 2^i \tag{5.12}$$

where $a \in \{-1, 0, 1\}$ [21]. In other words, the expression (5.12) allows negative coefficients as well. For example, the binary representation of the integer 29 is $29 = 2^0 + 2^2 + 2^3 + 2^4 = [10111]$. Therefore, its Hamming weight is 4. On the other hand, if negative coefficients are allowed, the integer 29 can be represented as $29 = 2^5 - 2^1 - 2^0$. Therefore, the arithmetic weight of 29 is 3. The arithmetic weight is always smaller than or equal to the number of 1 s in the binary representation.

The arithmetic distance between two integers N_1 and N_2 is defined by [21]

$$A_d(N_1, N_2) = wt(N_1 - N_2).$$

For example, $A_d(29, 33) = 14$ because $29 - 33 = -4 = -2^2$. Note, however, that $H_d(101110, 100001) = 3$. The arithmetic distance between two words is always smaller than or equal to their Hamming distance. Since one error in an arithmetic operation may cause a large Hamming distance between the erroneous and correct words, arithmetic distance is a more suitable measure for arithmetic codes.

As with Hamming distance, an arithmetic distance of at least d between every pair of codewords is necessary and sufficient for detecting all errors in $(d - 1)$ bits or less. For correcting c of fewer errors, a minimum arithmetic distance $2c + 1$ is necessary and sufficient. The correction is performed to the nearest codeword to the erroneous word in terms of arithmetic distance. Correction of any combination of c or fewer errors and simultaneous detection of a additional error requires the arithmetic distance $A_d \geq 2c + a + 1$ [21].

Two common types of arithmetic codes are AN codes and residue codes.

5.7.1 *AN* Codes

AN code is the simplest representative of arithmetic codes. The codewords are obtained by multiplying the integer N corresponding to the decimal representation of data by some constant A. For example, the defining table for $3N$ code is shown in Table 5.17.

To decode a codeword of an AN code, we divide it by A. If there is no remainder, no error has occurred. Otherwise, there is an error.

The reader may notice the analogy between AN and cyclic codes. In an AN code, a codeword is a multiple of A. In a cyclic code, a codeword polynomial generated by $g(x)$ is a multiple of $g(x)$.

Table 5.17 Defining table for a (4, 2) linear code

Data		Codeword			
d_0	d_1	c_0	c_1	c_2	c_3
0	0	0	0	0	0
0	1	0	1	1	0
1	0	1	1	0	0
1	1	1	0	0	1

A codeword of an AN code is by $\lceil \log_2 A \rceil$-bits longer than the encoded data word. So, the information ratio of AN codes for k-bit data is $k/(k + \lceil \log_2 A \rceil)$.

The AN code is an invariant with respect to addition and subtraction, but not to multiplication and division. For example, obviously $3(a \times b) \neq 3a \times 3b$ for $a, b \neq 0$.

The constant A determines the information rate of the code and its error detection capability. For binary codes, A should not be even. For, if A is even, then every codeword is divisible by two. Therefore, the lowest order bit of every codeword is always 0, i.e. useless.

In order to construct an AN code detecting all single errors, we need to ensure that the minimum arithmetic distance between every pair of codewords is 2. To guarantee that no two codewords are on arithmetic distance 1 apart, we need to select A such that

$$AN_1 - AN_2 = A(N_1 - N_2) \neq 2^i.$$

where N_1 and N_2 are integers corresponding to the decimal representation of data. The above condition can be satisfied by choosing A to be co-prime with 2 and greater than 2 [21]. The smallest A which satisfies it is $A = 3$. A $3N$ code detects all single errors.

It is more difficult to construct AN codes with arithmetic distance $A_d > 2$. The interested reader is referred to [21].

5.7.2 Residue Codes

Residue codes are separable arithmetic codes which are created by computing a *residue* for data and appending it to the data. The residue is generated by dividing the integer N corresponding to the decimal representation of data by an integer, called *modulus*.

Let N be the integer corresponding to the decimal representation of the binary data. The residue representation of N is obtained as

$$N = a \times m + r$$

where "×' and "+" are arithmetic multiplication and addition, respectively, and a, m and r are integers such as $0 \leq r \leq m$. The integer m is called the *modulus* and the integer r is called the *residue* of the operation $N \bmod m$. The $\lceil \log_2 m \rceil$-bit binary representation of r gives us the check bits of the resulting residue code.

For example, if $N = 11$ and $m = 8$, then $11 \bmod 8 = 3$. Since $\lceil \log_2 m \rceil = 3$, the 3-bit binary expression for check bits is [110]. By appending them to binary representation of data $[d_0 d_1 d_2 d_3] = [1101]$, we get the 7-bit codeword [1101110].

Again, the reader may notice the analogy between residue and separable cyclic codes. In a residue code with modulus m, the check bits are computed by dividing the data by the modulus. In a separable cyclic code, the polynomial representing check bits is generated by dividing the data polynomial by $g(x)$.

Residue codes are invariant with respect to addition, since

$$(b + c)\text{mod}m = b\text{mod}m + c\text{mod}m$$

where b and c are data words and m is modulus. This allows us to handle residues separately from data during the addition process. The value of the modulus determines the information rate and the error detection capability of the code.

An interesting class of residue codes is *low-cost residue codes* [24], which use modulus of type $m = 2^r - 1, r \geq 2$. Since

$$2^{r \times i}\text{mod}(2^r - 1) = 1$$

or any $i \geq 0$, the following property holds:

$$(d_i \times 2^{r \times i})\text{mod}(2^r - 1) = d_i\text{mod}(2^r - 1).$$

Therefore, the encoding for low-cost residue codes can be done by adding blocks of data of size r modulo m rather than dividing the data modulo r.

The check bits are computed by first partitioning the data into blocks of length r and then adding these blocks modulo $2^r - 1$. If the data length is not a multiple of r, the data are padded with 0s.

For example, suppose that the data are $[d_0 d_1 \ldots d_8] = [011011001]$ and $m = 3 = 2^2 - 1$. First, we pad the data with 0 to make its length a multiple of 2. We obtain $[d_0 d_1 \ldots d_8 d_9] = [0110110010]$. Note that padding does not change the value of the integer corresponding to the data, since we add 0 as the most significant bit. Then, we partition the data into blocks of length 2: [01], [10], [11], [00], [10]. Finally, we add these two blocks modulo 2. The result is [10], corresponding to 1 in decimal. To verify this result, we can divide the decimal representation 310 of the data by 3. We get $310\text{mod}3 = 1$. A we can see, both results are the same.

A variation of residue codes is *inverse residue codes*, where a Boolean complement of the residue, rather than the residue itself, is appended to the data. Such codes have been shown to be better than usual residue codes for detecting common-mode faults [2].

5.8 Summary

In this chapter, we have studied how information errors can be tolerated by means of coding. We have defined fundamental notions such as code, encoding and decoding, and information rate. We have considered many important families of codes, including parity codes. Linear codes, cyclic codes, unordered codes, and arithmetic codes.

Problems

5.1. What is the maximum length of data which can be encoded by a binary code of length 4 and size 8?

5.2. Give an example of a binary code of length 6 and size 8 which has the information rate 1/2.

5.3. Give an example of a binary code of length 5 and size 16 which has the information rate 4/5.

5.4. What is the main difference in the objectives of encoding for coding theory and encryption for cryptography?

5.5. What is the maximum Hamming distance between two words in the codespace $\{0, 1\}^4$?

5.6. Consider the code $C = \{000000, 010101, 101010, 111111\}$.

1. What is the code distance of C?
2. If C is used only for error detection, how many errors can it detect? Draw a gate-level logic circuit of an error-detection block which detects the maximum possible number of errors.
3. If C is used only for error correction, how many errors can it correct? Draw a gate-level logic circuit of a decoder which corrects the maximum possible number of errors.

5.7. Prove that it is possible to decode in a way that corrects all errors in c or fewer bits and simultaneously detects up to a additional errors if and only if the code distance is at least $2c + a + 1$.

5.8. M-ary codes can be used for protecting transmission and storage of m-ary digital data [7]. Suggest how the definitions of Hamming distance and code distance can be extended to ternary codes. Your definitions should preserve the following two properties: (1) to be able to correct c-digit errors, a ternary code should have a code distance of at least $2c + 1$; (2) to be able to detect d-digit errors, the ternary code should have a code distance of at least $d + 1$.

5.9. Prove that, for any $n > 1$, a parity code of length n has the code distance 2.

5.10.
1. Construct an even parity code for 4-bit data.
2. What is the information rate of this code?
3. Suppose that word [11010] is received. Assuming a single -bit error, which codewords have possibly been transmitted?

5.11. Draw a gate-level logic circuit of an odd parity generator for 5-bit data. Limit yourself to the use of two-input gates only.

5.12. Draw a gate-level logic circuit of an odd parity checker for 5-bit data. Limit yourself to the use of two-input gates only.

5.13.
1. Construct the horizontal and vertical even parity code for a 5×3 block of data.
2. Give an example of a 4-bit error which can be detected by the code.
3. Give an example of a 4-bit error which cannot be detected by the code.

4. Give an example of a 2-bit error which can be corrected by the code.
5. Give an example of a 2-bit error which cannot be corrected by the code.

5.14. How would you generalize the notion of parity for ternary codes? Give an example of a ternary parity code for 3-digit data which satisfies your definition.

5.15. Construct the generator matrix G and the parity check matrix H and for an even parity code for 6-bit data.

5.16. Construct the generator matrix G and the parity check matrix H and for a triplication code for 3-bit bit data.

5.17. Construct the parity check matrix H and the generator matrix G for a linear code for 5-bit data which can:

1. detect one error
2. correct one error
3. correct one error and detect one additional error.

5.18. Let H be a parity check matrix of an (n, k)-linear code. Prove that every pair of columns of H is linearly independent only if $n \leq 2^{n-k} - 1$.

5.19. Prove that, for any two vectors $v, u \in V^n$, if both v and u contain an odd number of 1's and $v \neq u$, then vector $v \oplus u$ contains an even number of 1's which is greater than 0.

5.20. Draw a gate-level logic circuit of a parity generator for $(7, 4)$ Hamming code with the generator matrix (5.10). Limit yourself to the use of two-input gates only.

5.21. 1. Construct the parity check matrix H and the generator matrix G of a Hamming code for 11-bit data.
2. Write logic equations for the check bits of the resulting code.

5.22. Construct the lexicographic parity check matrix H of a Hamming code for 11-bit data.

5.23. Construct the parity check matrix H and the generator matrix G of an extended Hamming code for 11-bit data.

5.24. Find the generator matrix for the (7,4) cyclic code C with the generator polynomial $1 \oplus x^2 \oplus x^3$. Prove that C is a Hamming code.

5.25. Find the generator matrix for the (15,11) cyclic code C with the generator polynomial $1 \oplus x \oplus x^4$. Prove that C is a Hamming code.

5.26. Compute the check polynomial for the (7,4) cyclic code with the generator polynomial $g(x) = 1 \oplus x^2 \oplus x^3$.

5.27. 1. Find a generator polynomial for the even parity code 3-bit data.
2. Find the check polynomial for this code.

5.28. Let C be and (n, k) cyclic code. Prove that the only burst errors of length $n - k + 1$ that are not detectable are shifts of scalar multiples of the generator polynomial.

5.29. Suppose you use a cyclic code generated by the polynomial $g(x) = 1 \oplus x \oplus x^3$. You have received a word $c(x) = 1 \oplus x \oplus x^4 \oplus x^5$. Check whether an error has occurred during transmission.

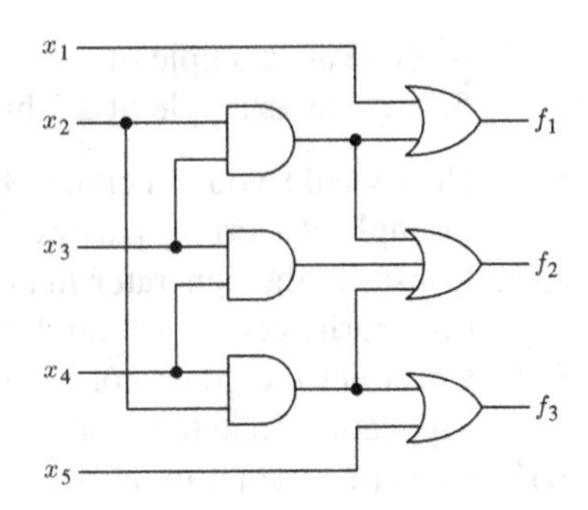

Fig. 5.15 Logic diagram of a circuit in which any single stuck-at fault causes a unidirectional error at the outputs

5.30. Construct an LFSR implementing the multiplication by the generator polynomial $g(x) = 1 \oplus x \oplus x^4$. Show the state table for the 4-bit data $d(x) = 1 \oplus x \oplus x^3$ (as the one in Table 5.8).

5.31. Construct an LFSR implementing the division by the generator polynomial $g(x) = 1 \oplus x \oplus x^4$. Show the state table for the 8-bit word $c(x) = 1 \oplus x^3 \oplus x^4 \oplus x^5 \oplus x^6 \oplus x^7$ (as the one in Table 5.9). Is $c(x)$ a valid codeword?

5.32. 1. Construct an LFSR for decoding for CRC codes with the following generator polynomials:

$$\begin{array}{ll} \text{CRC} - 16: & 1 \oplus x^2 \oplus x^{15} \oplus x^{16} \\ \text{CRC} - \text{CCITT}: & 1 \oplus x^5 \oplus x^{12} \oplus x^{16} \end{array}$$

You may use "..." between the registers 2 and 15 in the 1st polynomial and 5 and 12 in the second, to make the picture shorter.

2. Use the first generator polynomial for encoding the data $1 \oplus x^3 \oplus x^4$.
3. Suppose that the error $1 \oplus x \oplus x^2$ is added to the codeword you obtained in the previous task. Check if this error will be detected or not.

5.33. Consider the logic circuit shown in Fig. 5.15. Check if it is possible for a stuck-at fault at a single line in this circuit to cause a bi-directional error in the output vector $[f_1 f_2 f_3]$.

5.34. Construct a Berger code for 3-bit data. What code distance has the resulting code?

5.35. Prove that every pair of codewords in a Berger code is unordered.

5.36. Suppose we know that 4-bit data will never include the word [0000]. Can we reduce the number of check bits required in a Berger code and still detect all unidirectional errors?

5.37. 1. Construct a $3N$ arithmetic code for 3-bit data.

2. Give an example of a fault which is detected by such a code and an example of a fault which is not detected by such a code.

5.38. Consider the following code:

0 0 0 1 1 1
0 0 1 1 1 0
0 1 0 1 0 1
0 1 1 1 0 0
1 0 0 0 1 1
1 0 1 0 1 0
1 1 0 0 0 1
1 1 1 0 0 0

1. What kind of code is it?
2. Is it a separable code?
3. What is the code distance of C?
4. What kind of faults can it detect/correct?
5. Design a gate-level logic circuit for encoding of 3-bit data in this code. Your circuit should have three inputs for data bits and six outputs for codeword bits.
6. How would you suggest doing error detection for this code?

References

1. Anderson, D., Shanley, T.: Pentium Processor System Architecture, 2nd edn. Addison-Wesley, Reading (1995)
2. Avižienis, A.: Arithmetic error codes: cost and effectiveness studies for application in digital system design. IEEE Trans. Comput. **20**(11), 1322–1331 (1971)
3. Berger, J.M.: A note on an error detection code for asymmetric channels. Inform. Contr. **4**, 68–73 (1961)
4. Berlekamp, E.R.: Nonbinary BCH decoding. In: International Symposium on Information Theory, San Remo, Italy (1967)
5. Bryant, V.: Metric Spaces: Iteration and Application. Cambridge University Press, Cambridge (1985)
6. De Micheli, G., Brayton, R., Sangiovanni-Vincentelli, A.: Optimal state assignment for finite state machines. IEEE Trans. Comput. Aided Des. Integr. Circ. Syst. **4**(3), 269–285 (1985)
7. Dubrova, E.: Multiple-valued logic in VLSI: Challenges and opportunities. In: Proceedings of NORCHIP'99, pp. 340–350 (1999)
8. Dubrova, E., Mansouri, S.: A BDD-based approach to constructing LFSRs for parallel CRC encoding. In: Proceedings of International Symposium on Multiple-Valued Logic, pp. 128–133 (2012)
9. Golomb, S.: Shift Register Sequences. Aegean Park Press, Laguna Hills (1982)
10. Hamming, R.: Error detecting and error correcting codes. Bell Syst. Tech. J. **26**(2), 147–160 (1950)
11. Immink, K.: Reed-Solomon codes and the compact disc. In: S. Wicker, V. Bhargava (eds.) Reed-Solomon Codes and Their Applications, pp. 41–59. Wiley, New York (1999)
12. Intel Corporation: Embedded Pentium processor family developer's manual (1998) http://download.intel.com/design/intarch/manuals/27320401.pdf
13. Jha, N., Wang, S.J.: Design and synthesis of self-checking VLSI circuits. IEEE Trans. Comput. Aided Des. **12**, 878–887 (1993)

14. Johnson, B.W.: The Design and Analysis of Fault Tolerant Digital Systems. Addison-Wesley, Reading (1989)
15. Lala, P.: Self-Checking and Fault-Tolerant Digital Design. Morgan Kauffmann Publishers, Waltham (2001)
16. Lavagno, L., Nowick, S.: Asynchronous control circuits. In: S. Hassoun, T. Sasao (eds.) Logic Synthesis and Verification, Lecture Notes in Computer Science, pp. 255–284. Kluwer Academic Publishers, Boston (2002)
17. Lidl, R., Niederreiter, H.: Introduction to Finite Fields and their Applications. Cambridge University Press, Cambridge (1994)
18. Mitra, S., McCluskey, E.J.: Which concurrent error detection scheme to choose? In: Proceedings of the 2000 IEEE International Test Conference, pp. 985–994 (2000)
19. Murray, B.: Journey into space. NASA Historical Reference Collection, pp. 173–175 (1985)
20. Peterson, W., Brown, D.: Cyclic codes for error detection. Proc. IRE **49**(1), 228–235 (1961)
21. Peterson, W.W., Weldon, E.J.: Error-Correcting Codes, 2nd edn. MIT Press, London (1972)
22. Pierce, J.: Optical channels: practical limits with photon counting. IEEE Trans. Commun. **26**(12), 1819–1821 (1978)
23. Reed, I.S., Solomon, G.: Polynomial codes over certain finite fields. J. Soc. Ind. Appl. Math. **8**(2), 300–304 (1960)
24. Sayers, I., Kinniment, D.: Low-cost residue codes and their application to self-checking vlsi systems. IEE Proc. Comput. Digit. Tech. **132**(4), 197–202 (1985)
25. Shannon, C.E.: A mathematical theory of communication. Bell Syst. Tech. J. **27**, 379–423 (1948)
26. Smith, D.H., Hughes, L.A., Perkins, S.: A new table of constant weight codes of length greater than 28. Electron. J. Comb. **13**, A2 (2006)
27. Stavinov, E.: A practical parallel CRC generation method. Feature Article, pp. 38–45 (2010)
28. Strang, G.: Introduction to Linear Algebra, 4th edn. Wellesley-Cambridge Press, Wellesley (2009)
29. Tang, D.T., Woo, L.S.: Exhaustive test pattern generation with constant weight vectors. IEEE Trans. Comput. **C-32**, 1145–1150 (1983)
30. Tezzaron Semiconductor: Soft errors in electronic memory (2004). http://www.tezzaron.com/about/papers/papers.htm
31. United States Federal Standard 1037C: Telecommunications: Glossary of telecommunication terms. General Services Administration (1996)
32. United States Postal Service: Domestic mail manual 708.4 - special standards, technical specifications, barcoding standards for letters and flats (2010). http://pe.usps.com/cpim/ftp/manuals/dmm300/708.pdf
33. Wells, R.B.: Applied Coding and Information Theory for Engineers. Prentice-Hall, Englewood Cliffs (1998)

Chapter 6
Time Redundancy

"Failure is simply an opportunity to begin again, this time more intelligently."

Henry Ford

Hardware redundancy impacts the size, weight, power consumption, and cost of a system. In some applications, it is preferable to use extra time rather than extra hardware to tolerate faults. In this chapter, we describe time redundancy techniques for detection and correction of transient faults. We also show how time redundancy can be combined with some encoding scheme to handle permanent faults.

The chapter is organized as follows. We begin with transient fault detection and correction. Then, we discuss how time redundancy can be used for distinguishing between transient and permanent faults. Afterwards, we consider four approaches for detecting permanent faults: alternating logic, recomputing with shifted operands, recomputing with swapped operands, and recomputing with duplication with comparison.

6.1 Transient Faults

Time redundancy involves repeating the computation or data transmission two or more times and comparing results with previously stored copies.

If a fault is transient, then stored results differ from the recomputed one. If the repetition is done twice, as shown in Fig. 6.1, a fault can be detected. If the repetition is done three times, as shown in Fig. 6.2, a fault can be corrected. Voting techniques similar to the ones used for hardware redundancy can be used for selecting the correct result.

Time redundancy can also be useful for distinguishing between transient and permanent faults. If a fault disappears after recomputation, we can assume that it

E. Dubrova, *Fault-Tolerant Design*, DOI: 10.1007/978-1-4614-2113-9_6,

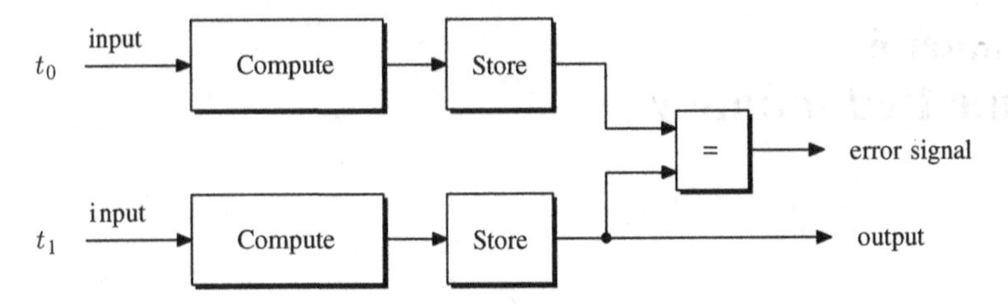

Fig. 6.1 Time redundancy for transient fault detection

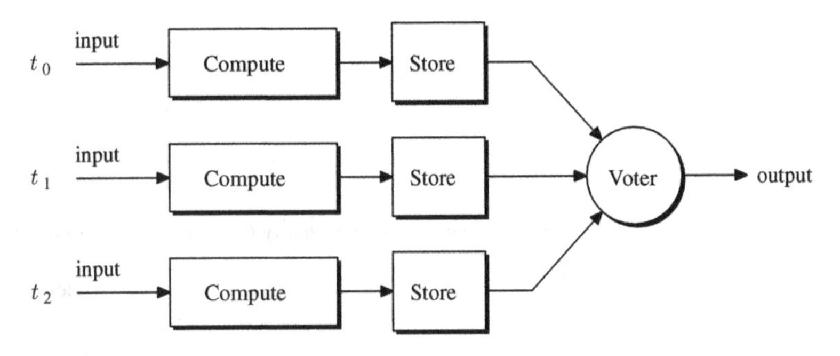

Fig. 6.2 Time redundancy for transient fault correction

is transient. In this case, there is no need to reconfigure a system and replace the affected hardware module. It can remain in operation, saving resources.

The main problem with time redundancy is the assumption that the data required to repeat a computation is available in the system [4]. Since a transient fault may cause a system failure, the computation may be difficult or not possible to repeat.

6.2 Permanent Faults

If time redundancy is combined with some encoding scheme, it can be used for detecting and correcting permanent faults. In this section, we present four time-redundancy techniques targeting permanent faults: alternating logic [7], recomputing with shifted operands [6], recomputing with swapped operands [2], and recomputing with duplication with comparison [3].

6.2.1 Alternating Logic

Alternating logic is a simple but efficient time-redundancy technique which can be used for the detection of permanent faults in data transmission and in logic circuits [7].

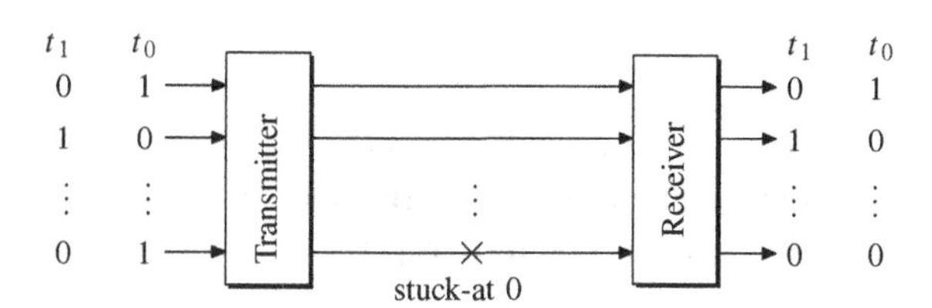

Fig. 6.3 Alternating logic technique

Data Transmission

Suppose the data is transmitted over a parallel bus, as shown in Fig. 6.3. At time t_0, the original data is transmitted. Then, at time t_1, the data is complemented and retransmitted. The two received words are compared to check if they are bit-wise complements of each other. Any disagreement indicates a fault. Such a scheme is capable of detecting all single and multiple stuck-at faults at the bus lines.

As an example, suppose that a single stuck-at-0 fault has occurred at the bus line marked by a cross in Fig. 6.3. Suppose that the data word $[10\ldots 1]$ is transmitted. At time t_0, we receive the word $[10\ldots 0]$. At time t_1, we receive the word $[01\ldots 0]$, in which all bits but the last are complements of the corresponding bits in the first word. So, we can conclude that a stuck-at-0 fault has occurred at the last bus line.

Example 6.1. Which type of encoding would you recommend to combine with the alternating logic redundancy to enable correction of all single stuck-at faults at the bus lines?

Single stuck-at faults can be corrected if we encode the transmitted data word $[d_0d_1\ldots d_{k-1}]$ using a parity code. For example, if even parity is used, then the resulting codeword is $[d_0d_1\ldots d_{k-1}p]$, where $p = d_0 \oplus d_1 \oplus \ldots \oplus d_{k-1}$. At time t_0, we submit the codeword $[d_0d_1\ldots d_{k-1}p]$. At time t_1, we submit its bit-wise complement $[\overline{d}_0\overline{d}_1\ldots \overline{d}_{k-1}\overline{p}]$.

Note that $\overline{p}$ is the complement of p rather than the parity bit for the word $[\overline{d}_0\overline{d}_1\ldots \overline{d}_{k-1}]$. For example, if k is even, then $\overline{p}$ makes the overall parity of $[\overline{d}_0\overline{d}_1\ldots \overline{d}_{k-1}\overline{p}]$ odd.

To check if a single stuck-at fault has occurred at the bus lines, we compare d_i with $\overline{d}_i$ for all $i \in \{0, 1, \ldots, k-1\}$ and p with $\overline{p}$. If there is a disagreement in some bit position, we recompute the parity bit for the word transmitted at time t_0. If the recomputed parity bit, p^*, differs from p, then the fault is in the word transmitted at time t_0. We correct the bit position in which the disagreement occurred. Otherwise, if $p^* = p$, then the fault occurred in the word transmitted at time t_1. Again, we correct the bit position in which the disagreement occurred.

Note that we will also be able to correct all multiple stuck-at faults in an odd number of bit positions. However, multiple faults in an even number of bit positions cannot be corrected in this way.

Logic Circuits

Alternating logic can also be used for detecting permanent faults in logic circuits implementing self-dual Boolean functions. A *dual* of a Boolean function $f(x_1, x_2, \ldots, x_n)$ is defined as [5]

$$f_d(x_1, x_2, \ldots, x_n) = \overline{f}(\overline{x}_1, \overline{x}_2, \ldots, \overline{x}_n).$$

For example, a 2-variable AND $f(x_1, x_2) = x_1 \cdot x_2$ is the dual of a 2-variable OR $(x_1, x_2) = x_1 + x_2$, and vice versa.

A Boolean function in said to be *self-dual* if it is equal to its dual, i.e.,

$$f(x_1, x_2, \ldots, x_n) = f_d(x_1, x_2, \ldots, x_n).$$

So, the value of a self-dual function f for the input assignment $(x_1, x_2, \ldots, x_n)$ equals the value of the complement of f for the input assignment $(\overline{x}_1, \overline{x}_2, \ldots, \overline{x}_n)$.

Examples of self-dual functions are sum, s, and carry-out, c_{out}, output functions of a full-adder. Table 6.1 shows their truth tables. It is easy to see that the property $f(x_1, x_2, \ldots, x_n) = \overline{f}(\overline{x}_1, \overline{x}_2, \ldots, \overline{x}_n)$ holds for both functions.

For a fault-free circuit implementing a self-dual Boolean function, input assignments $(x_1, x_2, \ldots, x_n)$ and $(\overline{x}_1, \overline{x}_2, \ldots, \overline{x}_n)$ produce output values which are complements of each other. Therefore, if a circuit has a fault, this fault can be detected by finding an input assignment for which $f(x_1, x_2, \ldots, x_n) = f(\overline{x}_1, \overline{x}_2, \ldots, \overline{x}_n)$.

For example, consider a circuit in Fig. 6.4 which implements the full-adder. The stuck-at-1 fault marked by a cross can be detected by applying the input assignment $(a, b, c_{\text{in}}) = (1, 0, 0)$ followed by its complement (0, 1, 1). If an adder is fault-free, the resulting output values will be $s(1, 0, 0) = 1$, $c_{\text{out}}(1, 0, 0) = 0$ and $s(0, 1, 1) = 0$, $c_{\text{out}}(0, 1, 1) = 1$. However, in the presence of the stuck-at-1 fault marked by a cross, we get $s(1, 0, 0) = 1$, $c_{\text{out}}(1, 0, 0) = 1$ and $s(0, 1, 1) = 0$, $c_{\text{out}}(0, 1, 1) = 1$. Since $c_{\text{out}}(1, 0, 0) = c_{\text{out}}(0, 1, 1)$, the fault is detected.

Any n-variable Boolean function $f(x_1, x_2, \ldots, x_n)$ which is not self-dual can be transformed to a self-dual Boolean function of $n + 1$-variables as follows:

Table 6.1 Truth table of a full-adder

a	b	c_{in}	s	c_{out}
0	0	0	0	0
0	0	1	1	0
0	1	0	1	0
0	1	1	0	1
1	0	0	1	0
1	0	1	0	1
1	1	0	0	1
1	1	1	1	1

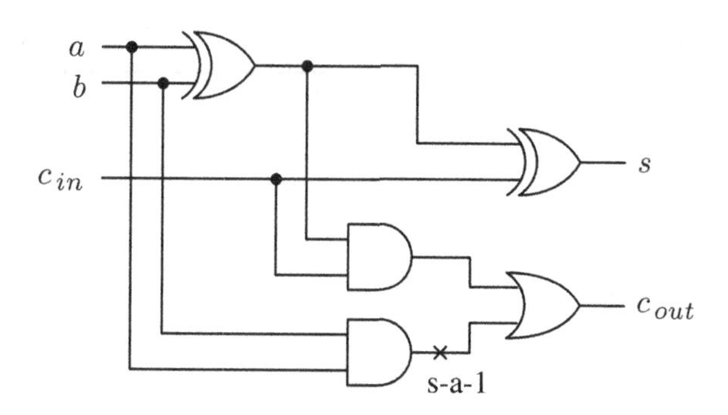

Fig. 6.4 Logic circuit implementing a full-adder with a single stuck-at fault

$$g_{sd}(x_1, x_2, \ldots, x_n, x_{n+1}) = x_{n+1} f(x_1, x_2, \ldots, x_n) + \overline{x}_{n+1} f_d(x_1, x_2, \ldots, x_n)$$

where f_d is the dual of f [1]. The new variable x_{n+1} is a control variable deciding if the value of f or f_d determines the value of g_{sd}. It is easy to see that such a function g_{sd} produces complemented values for complemented inputs. A drawback of this technique is that the circuit implementing g_{sd} may be twice as large as the circuit implementing f.

Example 6.2. Check if the output function of a 2-out-of-3 majority voter is self-dual.

To check if a function f is self-dual, we need to compute its dual f_d and compare it to f.

The output function of a 2-out-of-3 majority voter can be expressed as follows:

$$f(x_1, x_2, x_3) = x_1 x_2 + x_1 x_3 + x_2 x_3.$$

Next, we derive its dual function:

$$\begin{aligned} f_d(x_1, x_2, x_3) = \overline{f}(\overline{x}_1, \overline{x}_2, \overline{x}_3) &= (\overline{\overline{x}_1 \overline{x}_2 + \overline{x}_1 \overline{x}_3 + \overline{x}_2 \overline{x}_3}) \\ &= (x_1 + x_2)(x_1 + x_3)(x_2 + x_3) \\ &= (x_1 + x_2)(x_1 x_2 + x_1 x_3 + x_2 x_3 + x_3) \\ &= (x_1 + x_2)(x_1 x_2 + x_3) \\ &= x_1 x_2 + x_1 x_3 + x_2 x_3. \end{aligned}$$

Since $f = f_d$, the output function of a 2-out-of-3 majority voter is self-dual.

Example 6.3. 2-input XOR is not a self-dual function. Convert it to a 3-variable self-dual function.

First, we compute the dual of $f(x_1, x_2) = x_1 \oplus x_2$ as

$$f_d(x_1, x_2, x_3) = \overline{f}(\overline{x}_1, \overline{x}_2) = (\overline{\overline{x}_1 \oplus \overline{x}_2}) = (\overline{x_1 \oplus x_2}).$$

Table 6.2 Truth table for self-dual function constructed in Example 6.3

x_1	x_2	x_3	g_{sd}
0	0	0	1
0	0	1	0
0	1	0	0
0	1	1	1
1	0	0	0
1	0	1	1
1	1	0	1
1	1	1	0

Then, we transform XOR to a self-dual function of 3-variables as:

$$g_{sd}(x_1, x_2, x_3) = x_3(x_1 \oplus x_2) + \overline{x}_3(\overline{x_1 \oplus x_2}).$$

Table 6.2 shows the truth table of the resulting function g_{sd}. It is easy to see that the property $g_{sd}(x_1, x_2, x_3) = \overline{g}_{sd}(\overline{x}_1, \overline{x}_2, \overline{x}_3)$ holds.

6.2.2 Recomputing with Modified Operands

Several time-redundancy techniques for online fault detection in arithmetic logic units (ALUs) with bit-sliced organization have been proposed. Their basic idea is to modify the operands before performing the recomputation so that a permanent error affects different parts of the operands. For example, the operands may be shifted or their upper and lower parts may be swapped. In this section, we describe three such techniques: recomputing with shifted operands [6], recomputing with swapped operands [2], and recomputing with duplication with comparison [3].

Recomputing with Shifted Operands

In the recomputing with shifted operands, at time t_0, the computation is performed on the original operands. Then, at time t_1, the operands are shifted left, the computation is repeated on the shifted operands and then the result is shifted right. The results of two computations are compared. If they are equal, there is no error. Otherwise, an error has occurred. The shift can be either 1-bit (for logical operations) or 2-bits (for arithmetic operations) [6].

Figure 6.5 illustrates the recomputing with shifted operands technique for logical operations. A and B are two k-bit operands and R is the result of a logic operation on A and B. $X^<$ denotes a vector X shifted left. Suppose that a permanent fault has occurred in the bit slice i. At time t_0, this fault affects the bit r_i. At time t_1, A and B are shifted left by one bit and the computation is performed in the shifted operands.

Fig. 6.5 Recomputing with the shifted operands technique with 1-bit left shift for logical operations

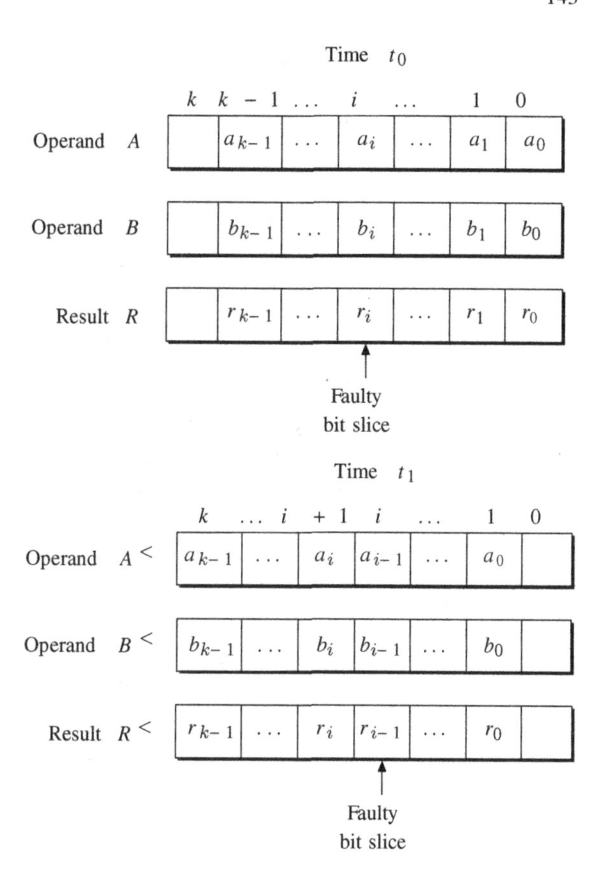

Therefore, the fault in the bits slice i affects r_{i-1}. When R is shifted right and the results of the two computations are compared, they disagree in either the r_i, or r_{i-1}, or in both bits, unless the fault did not cause an error in either result.

As we can see, a 1-bit left shift is capable of detecting permanent faults in any single bit slice.

Example 6.4. The recomputing with shifted operands is applied to perform the bit-wise XOR on 8-bit operands A and B. Suppose that the bit slice 3 is permanently stuck to 1 (bits are counted from the right starting from 0). Check in which bits the results of two computations disagree if $A = [01101011]$ and $B = [01111000]$.

A fault-free bit-wise XOR on A and B would have resulted in:

$$\begin{array}{r} 7\,6\,5\,4\,3\,2\,1\,0 \\ A = 0\,1\,1\,0\,1\,0\,1\,1 \\ B = 0\,1\,1\,1\,1\,0\,0\,0 \\ \hline R = 0\,0\,0\,1\,0\,0\,1\,1 \end{array}$$

Since the bit slice 3 is stuck to 1, we get $R_{t_0} = [00011011]$ instead.

During the second computation, we shift A and B 1 bit left and perform XOR on the resulting 9-bit words. If there had been no fault, we would have obtained:

$$
\begin{array}{r}
8\,7\,6\,5\,4\,3\,2\,1\,0 \\
A^{<} = 0\,1\,1\,0\,1\,0\,1\,1\,0 \\
B^{<} = 0\,1\,1\,1\,1\,0\,0\,0\,0 \\
\hline
R^{<} = 0\,0\,0\,1\,0\,0\,1\,1\,0
\end{array}
$$

Instead, we get [000101110]. After shifting this word 1 bit right, we get the 8-bit result $R_{t_1} = [00010111]$. We can see that R_{t_0} and R_{t_1} disagree in bits 2 and 3.

For arithmetic operations such as the addition using a ripple-carry adder, a 2-bit shift is required to guarantee the detection of permanent faults in any single bit slice. Figure 6.6 illustrates the case. A and B are two k-bit operands and S and C are k-bit

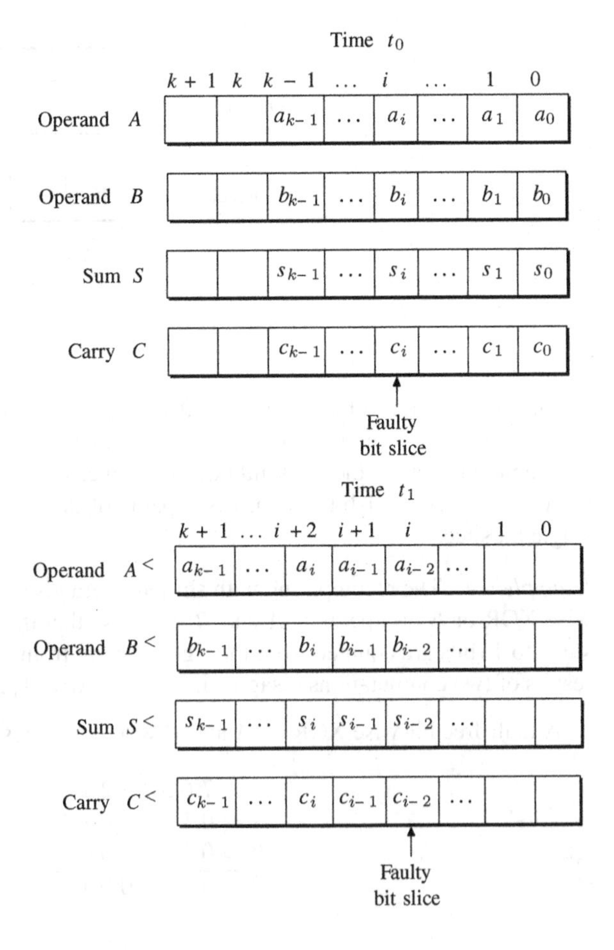

Fig. 6.6 Recomputing with shifted operands technique for arithmetic operations

sum and carry, respectively. Suppose that a permanent fault has occurred in the bit slice i. At time t_0, this fault affects the bits s_i and c_i is one of three possible ways:

1. The sum bit is erroneous. Then, the incorrect result differs form the correct one by either -2^i (if $s_i = 0$), or by 2^i (if $s_i = 1$).
2. The carry bit is erroneous. Then, the incorrect result differs from the correct one by either -2^{i+1} (if $c_i = 0$), or by 2^{i+1} (if $c_i = 1$).
3. Both sum and carry bits are erroneous. Then, there are four possibilities:

   - $s_i = 0$ and $c_i = 0$: -3×2^i;
   - $s_i = 0$ and $c_i = 1$: 2^i;
   - $s_i = 1$ and $c_i = 0$: -2^i;
   - $s_i = 1$ and $c_i = 1$: 3×2^i.

So, if the operands are not shifted, then the erroneous result differs from the correct result by one of the following values: $\{0, \pm 2^i, \pm 3 \times 2^i, \pm 2^{i+1}\}$.

At time t_1, A and B are shifted left by two bits and the addition is performed in the shifted operands. Therefore, the fault in the bits slice i affects s_{i-2} and c_{i-2} is one of three possible ways:

1. The sum bit is erroneous. Then, the incorrect result differs form the correct one by either -2^{i-2} (if $s_i = 0$), or by 2^{i-2} (if $s_i = 1$).
2. The carry bit is erroneous. Then, the incorrect result differs from the correct one by either -2^{i-1} (if $c_i = 0$), or by 2^{i-1} (if $c_i = 1$).
3. Both, sum and carry bits, are erroneous. Then, there are four possibilities:

   - $s_i = 0$ and $c_i = 0$: $-3 \times 2^{i-2}$;
   - $s_i = 0$ and $c_i = 1$: 2^{i-2};
   - $s_i = 1$ and $c_i = 0$: -2^{i-2};
   - $s_i = 1$ and $c_i = 1$: $3 \times 2^{i-2}$.

So, if the operands are shifted left by two bits, then the erroneous result differs from the correct result by one of the following values: $\{0, \pm 2^{i-2}, \pm 3 \times 2^{i-2}, \pm 2^{i-1}\}$.

We can see that the sets of values in which the erroneous results differ from the correct ones in both computations have only one element in common. This element is 0. Therefore, when S and C are shifted two bits right and the two results are compared, they differ unless they are both correct.

Now, let us investigate why a 1-bit left shift is not sufficient to guarantee the detection of all faults in the addition. Suppose that instead of shifting left by 2 bits, we shift left by only 1 bit. Suppose that at time t_1 the carry bit is erroneous. Then, the fault in the bit slice i will cause the value of the bit c_{i-1} to change either by -2^i (if $c_i = 0$), or by 2^i (if $c_i = 1$). However, in the computation performed at time t_0, the erroneous result may also differ from the correct one by $\pm 2^i$. This will happen if the sum bit s_i is erroneous. So, when S and C are shifted one bit right and the two results are compared, they are the same although they are not correct. Therefore, such a fault is not detected.

Example 6.5. A company producing ALUs with a bit-sliced organization decides to employ recomputing with shifted operands for permanent fault detection in arithmetic operations. However, the designer misunderstands the requirements and implements a 1-bit left shift instead of a 2-bit left shift. Suppose that the operands are 8-bit and the bit slice 3 is permanently stuck to 0. Check if this fault will be detected during the addition of the operands $A = [01100011]$ and $B = [00010010]$. Assume that a ripple-carry adder is used.

At time t_0, a fault-free addition would have resulted in:

$$\begin{array}{r} 7\,6\,5\,4\,3\,2\,1\,0 \\ A = 0\,1\,1\,0\,0\,0\,1\,1 \\ B = 0\,0\,0\,1\,0\,0\,1\,0 \\ \hline C = 0\,0\,0\,0\,0\,0\,1\,0 \\ S = 0\,1\,1\,1\,0\,1\,0\,1 \end{array}$$

Since the bit slice 2 is stuck to 0, we get $S_{t_0} = [01110001]$ instead. The second bit of the fault-free C is 0, therefore C is unchanged.

During the second computation, we shift A and B one bit left and add the resulting 9-bit vectors. If there had been no fault, we would have obtained:

$$\begin{array}{r} 8\,7\,6\,5\,4\,3\,2\,1\,0 \\ A^{<} = 0\,1\,1\,0\,0\,0\,1\,1\,0 \\ B^{<} = 0\,0\,0\,1\,0\,0\,1\,0\,0 \\ \hline C^{<} = 0\,0\,0\,0\,0\,0\,1\,0\,0 \\ S^{<} = 0\,1\,1\,1\,0\,1\,0\,1\,0 \end{array}$$

However, the fault in the bit slice 2 changes the carry to $C^{<}_{t_1} = [000000000]$. As a result, the sum changes to $S^{<}_{t_1} = [011100010]$. After shifting the sum vector 1 bit right, we get $S_{t_1} = [01110001]$. We can see that $S_{t_0} = S_{t_1}$. So, the fault will not be detected.

The recomputing with shifted technique requires the following extra hardware: three shift registers (to shift both operands left and to shift the result of the second computation right), a storage register for the result of the first computation, the comparator, and the additional bits in the ALU required for the shift operation. Furthermore, the faults which might occur in the extra registers or comparator will not be detected. The technique presented in the next section attempts to reduce the amount of extra hardware required for modifying the operands.

Recomputing with Swapped Operands

In recomputing with swapped operands, both operands are split into two halves. During the first computation, the operands are treated as usual. The second computation is performed with the lower and the upper halves of operands swapped. When the

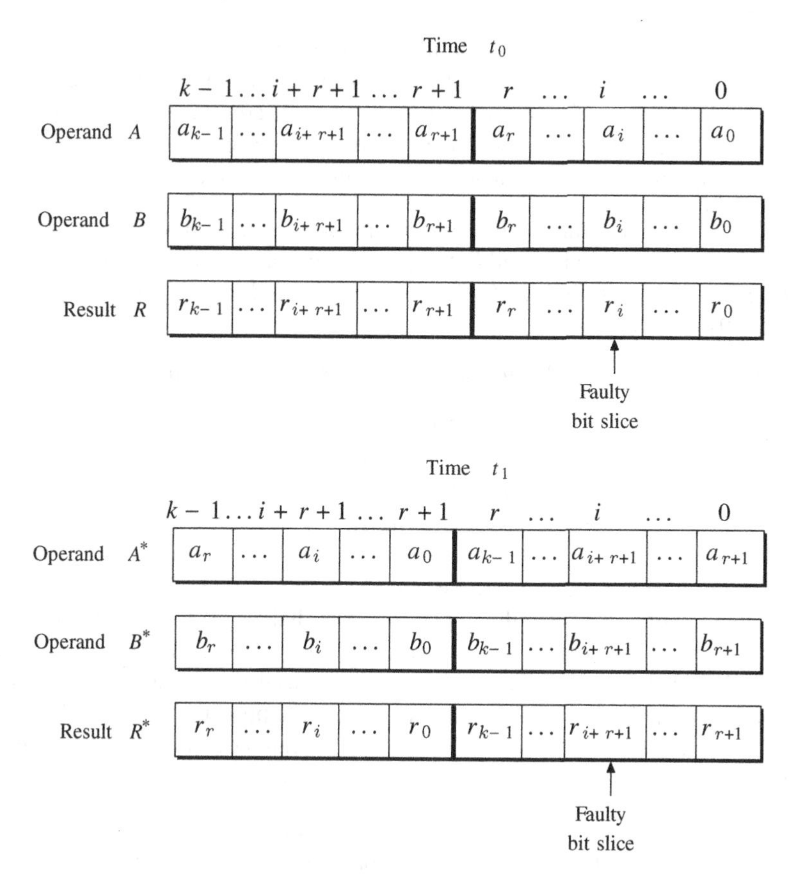

Fig. 6.7 Recomputing swapped operands technique for logical operations

computation is completed, the lower and the upper halves of the result are swapped back. The results of the two computations are compared. A disagreement indicates a fault [2].

Recomputing with swapped operands technique can detect permanent faults in any single bit slice in both logical and arithmetic operations. Figure 6.7 illustrates the case of logical operations. A and B are two k-bit operands and R is the result of a logic operation on A and B, where k is even. X^* denotes a vector X with the upper and the lower halves swapped. Suppose the lower half of an operand contains the bits from 0 to r and the upper half contains the bits from $r+1$ to $k-1$, for $r = k/2 - 1$. Suppose that a permanent fault has occurred in some bit slice i in the lower half of the operands, $i \in \{0, 1, \dots, r\}$. During the first computation, this fault affects the bit r_i. At time t_1, the operand halves are swapped and the computation is performed on operands with swapped halves. Therefore, the fault in the bits slice i affects the bit r_{i+r+1}. When the lower and the upper halves of the result R are swapped back and the results of the two computations are compared, they disagree in either the r_i, or r_{i+r+1}, or both bits, unless a fault did not cause an error in both results.

If a fault occurs in the bit slice i in the upper half of the operands, i.e., $i \in \{r+1, r+2, \ldots, k-1\}$, then during the second computation this fault affects the bit r_{i-r-1}. Therefore, when R is shifted right and R and R^* are compared, they disagree in either the r_i, or r_{i-r-1}, or both bits, unless they are correct.

As we can see, in both cases, a permanent fault in any single bit slice is detected.

Example 6.6. Recomputing with swapped operands is applied to perform the bit-wise AND on 8-bit operands A and B. Suppose that the bit slice 4 is permanently stuck to 1. Check in which bits the results of two computations disagree if $A = [11100111]$ and $B = [01101111]$.

A fault-free bit-wise AND on A and B would have resulted in

$$\begin{array}{r} 7\,6\,5\,4\,3\,2\,1\,0 \\ A = 1\,1\,1\,0\,0\,1\,1\,1 \\ B = 0\,1\,1\,0\,1\,1\,1\,1 \\ \hline R = 0\,1\,1\,0\,0\,1\,1\,1 \end{array}$$

Since the bit slice 4 is stuck to 1, we get $R_{t_0} = [01110111]$ instead. During the second computation, we swap the upper and lower halves of A and B and perform AND on the resulting vectors. If there had been no fault, we would have obtained $R^*_{t_1} = [01110110]$. Since the fault in the bit slice 4 has the same value as the correct value of the bit r_0, this fault causes no error. After swapping the upper and the lower halves back, we get the result $R_{t_1} = [01100111]$. We can see that R_{t_0} and R_{t_1} disagree in bit 4.

The recomputing with swapped operands technique can also detect permanent faults in any single bit slice during arithmetic operations. Consider a bit-sliced ripple-carry adder with k-bit operands where k is even. Suppose that a permanent fault has occurred in the bit slice i. By carrying out an analysis similar to the recomputing with shifted operands case, we can determine that, at time t_0, the erroneous result differs from the correct result by one of the following values: $\{0, \pm 2^i, \pm 3 \times 2^i, \pm 2^{i+1}\}$.

During time t_1, the addition of the operands is performed with the upper and lower halves swapped. Multiplexers are used to switch carry bits between the lower and the upper parts of the adder. Suppose that the lower half of an operand contains the bits from 0 to r and the upper half contains the bits from $r+1$ to $k-1$, for $r = k/2 - 1$. If $i \leq r$, then the fault in the bit slice i affects the bits s_{i+r+1} and c_{i+r+1} in one of three possible ways:

1. The sum bit is erroneous. Then, the incorrect result differs from the correct one by either -2^{i+r+1} (if $s_i = 0$), or by 2^{i+r+1} (if $s_i = 1$).
2. The carry bit is erroneous. Then, the incorrect result differs from the correct one by either -2^{i+r+2} (if $c_i = 0$), or by 2^{i+r+2} (if $c_i = 1$).
3. Both sum and carry bits are erroneous. Then, there are four possibilities:
   - $s_i = 0$ and $c_i = 0$: $-3 \times 2^{i+r+1}$;
   - $s_i = 0$ and $c_i = 1$: 2^{i+r+1};

- $s_i = 1$ and $c_i = 0$: -2^{i+r+1};
- $s_i = 1$ and $c_i = 1$: $3 \times 2^{i+r+1}$.

So, if the upper and lower halves of the operands are swapped, then the erroneous result differs from the correct result by one of the following values: $\{0, \pm 2^{i+r+1}, \pm 3 \times 2^{i+r+1}, \pm 2^{i+r+2}\}$.

Similarly, we can show that if $i > r$, then the erroneous result differs from the correct result by one of the following values: $\{0, \pm 2^{i-r-1}, \pm 3 \times 2^{i-r-1}, \pm 2^{i-r}\}$.

We can see that the sets of values in which the erroneous results differ from the correct ones have only the element 0 in common. Therefore, when the upper and the lower halves of S are swapped back and the two sums are compared, they differ unless they are both correct.

Example 6.7. Recomputing with swapped operands is applied to protect a ripple-carry adder from permanent faults. Check in which bits the results of two additions disagree if the operands and the fault are the same as in Example 6.5.

At time t_0, we get the same result as in Example 6.5, namely $S_{t_0} = [01110001]$.

During the second computation, we swap the upper and lower halves of A and B and perform the addition on the resulting vectors. A fault-free addition would have resulted in:

$$
\begin{array}{rl}
 & 7\,6\,5\,4\,3\,2\,1\,0 \\
A^* = & 0\,0\,1\,1\,0\,1\,1\,0 \\
B^* = & 0\,0\,1\,0\,0\,0\,0\,1 \\
\hline
C^* = & 0\,0\,1\,0\,0\,0\,0\,0 \\
S^* = & 0\,1\,0\,1\,0\,1\,1\,1
\end{array}
$$

However, the fault in bit slice 2 changes the sum to $S^*_{t_1} = [01010011]$. After swapping its upper and lower halves back, we get $S_{t_1} = [00110101]$. Since S_{t_0} and S_{t_1} disagree, the fault is detected.

Recomputing with swapped operands requires the following extra hardware: multiplexers to perform swapping of upper and lower halves of the operands and to switch carry bits between the lower and the upper parts of the adder when the operand halves are swapped, a storage register for the result of the first computation, and the comparator. Since multiplexers are simpler than shift registers, the hardware overhead of recomputing with swapped operands is smaller than that of recomputing with shifted operands. The technique presented in the next section reduces the hardware overhead even further.

Recomputing Using Duplication with Comparison

In recomputing using duplication with comparison, the operands are split into two halves. First, the operation is carried out on the duplicated lower halves and the results are compared. One of the results is stored to represent the lower half of the

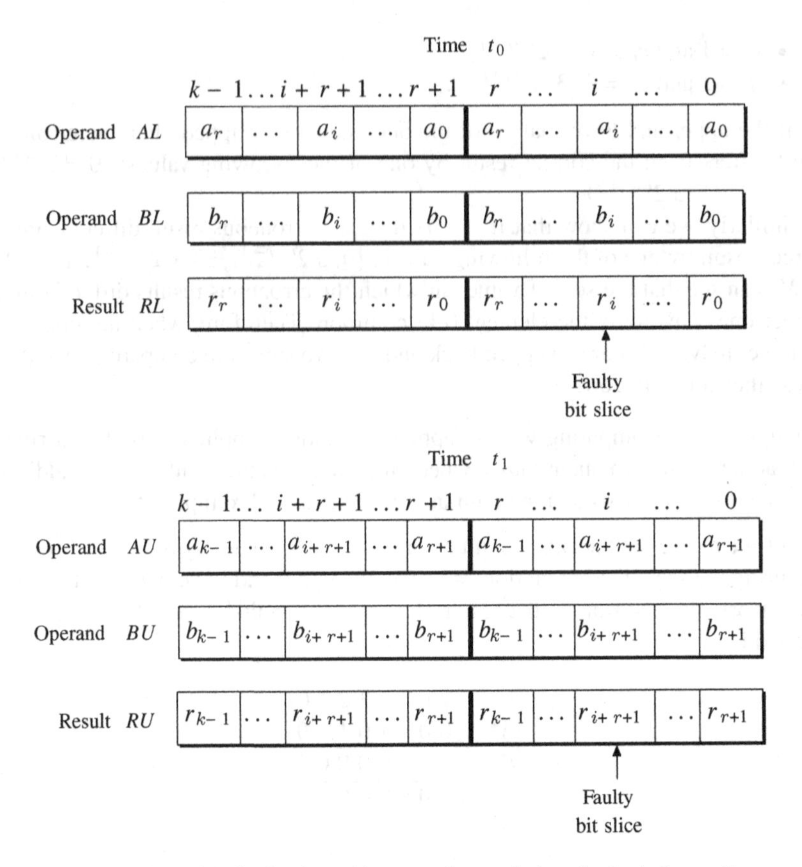

Fig. 6.8 Recomputing using duplication with comparison technique for logical operations

final result. Then, the same steps are repeated for the upper halves of the operands. Such a technique allows the detection of permanent faults in any single bit slice in both logical and arithmetic operations [3].

Figure 6.8 illustrates the case of logical operations. A and B are two k-bit operands and R is the result of a logic operation on A and B, where k is even. Let $AL(AU)$ and $BL(BU)$ denote the lower (upper) halves of A and B, respectively. Suppose the lower half of an operand contains the bits from 0 to r and the upper half contains the bits from $r+1$ to $k-1$, for $r = k/2 - 1$. Suppose that a permanent fault has occurred in some bit slice $i \in \{0, 1, \ldots, k-1\}$. During the first computation, this fault affects the bit r_i in one of the lower halves of the result. By comparing the two lower halves, this fault is detected.

At time t_1, the operation is repeated for the duplicated upper halves of the operands. The fault in the bits slice i affects the bit r_{i+r+1} in one of the upper halves of the result. By comparing the two upper halves, this fault is detected. We can see that in this way, we can detect any fault in a single bit slice.

Example 6.8. Recomputing using duplication with comparison is applied to perform the bit-wise OR on k-bit operands A and B. Suppose that the bit slices 0 and $k/2-1$ are permanently stuck to 0. Will such a fault be detected for any A, B, and k, for an even k? Illustrate your answer on the example of $A = [11100111]$, $B = [01101111]$ and $k = 8$.

Any fault which affects multiple bit slices in either the lower half, or the upper half of the final result (but not both) is detected by the recomputing using duplication with comparison technique. Since the bit slices 0 and $k/2-1$ both belong to the lower half of the final result, the fault affecting these two bit slices is detected.

For example, consider the case of $A = [11100111]$, $B = [01101111]$ and $k = 8$. Suppose that the bit slices 0 and 3 are permanently stuck to 0. During the first computation, bit-wise OR is performed on the duplicated lower halves of A and B. A fault-free operation would have resulted in:

$$\begin{array}{rl} & 7\ 6\ 5\ 4\ 3\ 2\ 1\ 0 \\ AL = & [0\ 1\ 1\ 1][0\ 1\ 1\ 1] \\ BL = & [1\ 1\ 1\ 1][1\ 1\ 1\ 1] \\ \hline RL = & [1\ 1\ 1\ 1][1\ 1\ 1\ 1] \end{array}$$

The stuck-at-0 fault in the bit slices 0 and 3 changes the result to [1111][0110]. By comparing [1111] with [0110], we find the disagreement and detect the fault.

Similarly, during the second computation, the bit-wise OR is performed on the duplicated upper halves of A and B. A fault-free operation would have resulted in:

$$\begin{array}{rl} & 7\ 6\ 5\ 4\ 3\ 2\ 1\ 0 \\ AU = & [1\ 1\ 1\ 0][1\ 1\ 1\ 0] \\ BU = & [0\ 1\ 1\ 0][0\ 1\ 1\ 0] \\ \hline RU = & [1\ 1\ 1\ 0][1\ 1\ 1\ 0] \end{array}$$

The stuck-at-0 fault in the bit slices 0 and 3 changes the result to [1110][0110]. By comparing [1110] with [0110] we find the disagreement and detect the fault.

In a similar way, recomputing using duplication with comparison can be applied to arithmetic operations. For example, suppose that the addition is performed using a ripple-carry adder. First, the lower and upper halves of the adder are used to compute the sum of the lower halves of the operands. Multiplexers are used to select the appropriate half of the operands and to handle the carries at the boundaries of the adder. The results are compared and one of them is stored to represent the lower half of the final sum. The second addition is carried out on the upper halves of the operands. Again, the lower and upper halves of the adder are used to compute the sum. The output carry from the first addition is added to both halves of the adder. The results are compared and one of them is used to represent the upper half of the final sum.

Suppose that a single fault has occurred in the bit slice i. During the first computation, this fault affects only one of the lower halves of the final sum. By comparing the two lower halves of the final sum, the fault is detected. Similarly, during the second computation, the fault affects only one of the upper halves of the final sum. By comparing the two upper halves of the final sum, the fault is detected.

Example 6.9. Recomputing using duplication with comparison is used to protect a k-bit full-adder from permanent faults, where k is even. Suppose that the bits slices i and $i + k/2 + 1$ have a permanent fault, for $i \in \{0, 1, \ldots, k/2 - 1\}$. Will such a fault be detected for any A, B, and k? Illustrate your answer in the example of $A = [10110010]$, $B = [00011011]$ and $k = 8$.

If a carry bit in the bit slice i is faulty, the fault may propagate to the sum bit in the bit slice $i + 1$. Since the fault in the bit slice $i + k/2 + 1$ affects the sum bit in the slice $k/2 + 1$, both halves of the final sum may potentially be affected in the same way. Therefore, if the bits slices i and $i + k/2 + 1$ are both faulty, such a fault may not be detected.

For example, consider the case of $A = [10110010]$, $B = [00011011]$ and $k = 8$. Suppose that the bit slices 1 and 6 are permanently stuck to 0. During the first computation, the duplicated lower halves of A and B are added. A fault-free addition would have resulted in:

$$\begin{array}{r l} & 7\ 6\ 5\ 4\ 3\ 2\ 1\ 0 \\ AL = & [0\ 0\ 1\ 0][0\ 0\ 1\ 0] \\ BL = & [1\ 0\ 1\ 1][1\ 0\ 1\ 1] \\ \hline CL = & [0\ 0\ 1\ 0][0\ 0\ 1\ 0] \\ SL = & [1\ 1\ 0\ 1][1\ 1\ 0\ 1] \end{array}$$

Due to the stuck-at-0 faults in the bit slices 1 and 6, we get instead:

$$\begin{array}{r l} CL_{t0} = & [0\ 0\ 1\ 0][0\ 0\ 0\ 0] \\ SL_{t0} = & [1\ 0\ 0\ 1][1\ 0\ 0\ 1] \end{array}$$

Since the two halves of SL_{t0} are the same, the fault is not detected.

During the second computation, the duplicated uppers halves of A and B are added. A fault-free addition would have resulted in:

$$\begin{array}{r l} & 7\ 6\ 5\ 4\ 3\ 2\ 1\ 0 \\ AU = & [1\ 0\ 1\ 1][1\ 0\ 1\ 1] \\ BU = & [0\ 0\ 0\ 1][0\ 0\ 0\ 1] \\ \hline CU = & [1\ 0\ 1\ 1][1\ 0\ 1\ 1] \\ SU = & [0\ 1\ 0\ 0][0\ 1\ 0\ 0] \end{array}$$

Due to the stuck-at-0 faults in the bit slices 1 and 6, we get instead:

$$\begin{array}{r l} CU_{t1} = & [1\ 0\ 1\ 1][1\ 0\ 0\ 1] \\ SU_{t1} = & [0\ 0\ 0\ 0][0\ 0\ 0\ 0] \end{array}$$

Since the two halves of SU_{t1} are the same, the fault is not detected.

Note that the same comparator is used twice to compare the two halves of the sum and the carry out from the most significant bit of the adder for errors. This comparator is nearly twice as small as the ones used in recomputing with shifted operands and recomputing with swapped operands techniques. Similarly, the storage register stores only one half of the result. Therefore, they are twice as small as the storage registers used in the other two techniques. Overall, the hardware overhead of recomputing using duplication with comparison technique is more than twice as small as that of the other two techniques [4].

6.3 Summary

In this chapter, we have studied time-redundancy techniques which can be used for detection or correction of transient and permanent faults. We have discussed the advantages and disadvantages associated with each approach.

Problems

6.1. Give three examples of applications where time is less important than hardware.

6.2. Check if the technique described in Example 6.1. It can be used for correcting transient faults. Give an example to illustrate your answer.

6.3. Check if the Boolean function $f(x_1, x_2, x_3) = x_1x_2 + \overline{x}_3$ is self-dual. Show its truth table to justify your answer.

6.4. In Example 6.2 we have shown that the output function of a 2-out-of-3 majority voter is self-dual. Figure 6.9 shows a logic circuit implementing a 2-out-of-3 majority voter. Find an input assignment which detects the stuck-at-1 fault marked by a cross.

6.5. Check if the output function of a 3-out-of-5 majority voter is self-dual.

6.6. Convert the 2-input AND to a 3-variable self-dual function.

6.7. Convert the 3-input XOR to a 4-variable self-dual function.

6.8. Recomputing with shifted operands is used to protect a k-bit ripple-carry adder from permanent faults. A 2-bit left shift is used. Suppose that the bits slices i

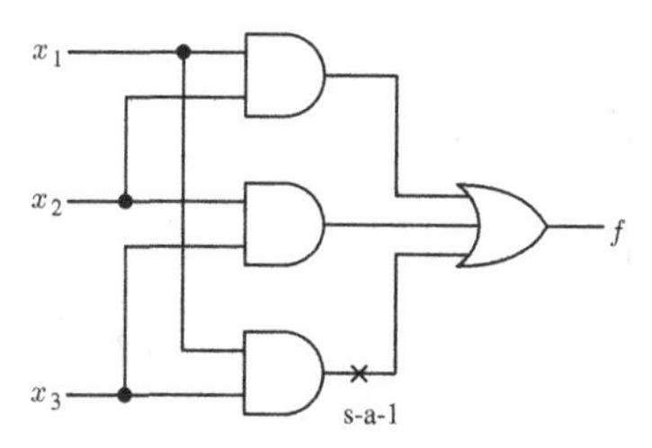

Fig. 6.9 Logic circuit implementing a 2-out-of-3 majority voter with a stuck-at fault

and $i + k/2 + 1$ have a permanent fault, for $i \in \{0, 1, \ldots, k/2 - 1\}$. Will such a fault be detected for any A, B, and k? Illustrate your answer in the example of some operands A and B and $k = 8$.

6.9. Repeat Problem 6.8 for the case when the fault affects two adjacent bits i and $i + 1$, for $i \in \{0, 1, \ldots, k - 2\}$.

6.10. Repeat Problem 6.8 for recomputing with swapped operands.

6.11. Repeat Problem 6.9 for recomputing with swapped operands.

6.12. Recomputing with shifted operands is applied to perform the bit-wise OR on k-bit operands A and B. Suppose that the bit slices 0 and $k/2 - 1$ are permanently stuck to 0. Will such a fault be detected for any A, B, and k? Illustrate your answer in the example of some operands A and B and $k = 8$.

6.13. Repeat Problem 6.12 for recomputing with swapped operands.

6.14. Check if recomputing with shifted operands will be able to detect permanent faults in any single bit slice of a k-bit ripple-carry adder if a 3-bit left shift is used instead of a 2-bit left shift. Justify your answer using the same reasoning we used for the case of a 2-bit left shift.

6.15. Check if recomputing with shifted operands will be able to detect permanent faults in any single bit slice for logical operations if a 1-bit right shift is used instead of a 1-bit left shift. Justify your answer using the same reasoning we used for the case of a 1-bit left shift.

6.16. Recomputing using duplication with comparison is applied to perform the bit-wise XOR on k-bit operands A and B, where k is even. Suppose that the bit slices 0 and $k/2$ are permanently stuck to 1. Will such a fault be detected for any A, B, and k? Illustrate your answer in the example of $A = [10100111]$, $B = [01101110]$ and $k = 8$.

6.17. Suppose that we modified the recomputing using duplication with comparison technique to recomputing using triplication with comparison as follows. The k-bit operands are split into three equal parts (assume that k is a multiple of 3). The operation is repeated three times on the triplicated parts and a majority voter is used to determine the correct result. Will such a technique be able to correct permanent faults in any single bit slice for: (a) logical operations and (b) arithmetic operations such as addition using a ripple-carry adder?

6.18. Suggest how you would modify the recomputing swapped operand technique to be able to correct permanent faults in any single bit slice for: (a) logical operations and (b) the addition operation, assuming that a ripple-carry adder will be used?

References

1. Biswas, N.N.: Logic Design Theory. Prentice Hall, Upper Saddle River (1993)
2. Johnson, B.: Fault-tolerant microprocessor-based systems. IEEE Micro **4**(6), 6–21 (1984)
3. Johnson, B., Aylor, J., Hana, H.: Efficient use of time and hardware redundancy for concurrent error detection in a 32-bit VLSI adder. IEEE J. Solid-State Circuits **23**(1), 208–215 (1988)

4. Johnson, B.W.: The Design and Analysis of Fault Tolerant Digital Systems. Addison-Wesley, New York (1989)
5. Kohavi, Z.: Switching and Authomata Theory, 2nd edn. McGraw-Hill, New York (1978)
6. Patel, J., Fung, L.: Concurrent error detection in ALU's by recomputing with shifted operands. IEEE Trans. Comput. **C-31**(7), 589–595 (1982)
7. Reynolds, D., Metze, G.: Fault detection capabilities of alternating logic. IEEE Trans. Comput. **C-27**(12), 1093–1098 (1978)

Chapter 7
Software Redundancy

"Programs are really not much more than the programmer's best guess about what a system should do."

Russel Abbot

Software fault tolerance techniques can be divided into two groups: single version and multi version [36]. Single-version techniques aim to improve the fault tolerance of a software component by adding to it mechanisms for fault detection, containment, and recovery. Multi-version techniques use redundant software components which are developed following design diversity rules.

As in the hardware case, various choices have to be examined to determine at which level the redundancy has to be provided and which modules are to be made redundant. The redundancy can be applied to a procedure, a process, or the whole software system. Usually, the components which have the lowest reliability are made redundant. One has to be aware that the increase in complexity caused by redundancy can be quite severe and may diminish the dependability improvement, unless redundant resources are allocated in a proper way.

The chapter is organized as follows. First, we analyze differences between hardware and software systems. Then, we describe common single-version techniques for fault detection, fault containment, and fault recovery in software. Afterwards, we study major multi-version techniques: recovery blocks, n-version programming, and n self-checking programming. We continue with a discussion of the importance of design diversity. Finally, we briefly cover software testing and common test coverage metrics.

7.1 Software Versus Hardware

In general, fault tolerance in the software domain is not as well understood and mature as fault tolerance in the hardware domain. In the case of hardware, we simplify the

E. Dubrova, *Fault-Tolerant Design*, DOI: 10.1007/978-1-4614-2113-9_7,

reliability evaluation by assuming that failures of components are independent events. It is usually not realistic to use such an assumption for software, where modules tend to have highly correlated failures [12]. The computation carried out in one module is normally directly or indirectly related to the computations performed by other modules. Therefore, an error in the results of one module affects the results of other modules.

There are controversial opinions on whether reliability can be used as a measure of the dependability of software. Software does not degrade with time. Its failures are mostly due to the activation of specification or design faults by the input sequences. So, if a fault exists in software, it will manifest itself the first time that the relevant conditions occur. This makes the reliability of a software module dependent on the environment that generates the input to the module over time. Different environments might result in different reliability values. The Ariane 5 rocket accident is an example of how a piece of software, safe for the Ariane 4 operating environment, can cause a disaster in a new environment [30].

Many current techniques for software fault tolerance resemble hardware redundancy schemes. For example, N-version programming is similar to N-modular hardware redundancy. N self-checking programming comes close to hot standby redundancy. However, software is inherently different from hardware. The traditional hardware fault tolerance techniques were developed to mitigate primarily permanent component faults, and secondarily transient faults caused by environmental factors [3]. They do not offer sufficient protection against design faults, which are dominant in software [33]. Obviously, we cannot tolerate a fault in a software module by triplicating it and voting results, because all copies have identical faults. Instead, each of the redundant modules has to be re-implemented in a different way.

Software is considerably more complex than hardware. For example, a collision avoidance system required on most commercial aircraft in the United States has 1,040 states [23]. The large number of states would not be a problem if the states exhibited regularity which is common in hardware. Unfortunately, software is quite irregular. Therefore, the number of states in it cannot be reduced by grouping the states into equivalence classes. This implies that only a very small part of the software system can be verified for correctness. Even the best quality software systems experience 3.3 faults per 1,000 lines of uncommented code [38]. Traditional testing and debugging methods [22] are inherently slow and unscalable. They are generally incapable of covering functional corner cases or finding hard-to-find bugs that may occur only after hundreds of thousands of cycles (like the Intel Pentium FDIV bug [41]). Formal methods promise higher coverage; however, due to their large computational complexity they are only applicable to specific applications [50]. As a consequence of incomplete verification, many design faults in software remain undetected, creating a risk of serious accidents like the Therac-25 radiation therapy machine overdoses [28], the sinking of the British destroyer Sheffield by a French-built Exocet missile after it was mistaken for one of their own [29], the explosion of the Ariane 5 rocket [30], and the breakup of the space shuttle Challenger [14].

7.2 Single-Version Techniques

Single-version techniques add to a software component, a number of functional capabilities that are unnecessary in a fault-free environment. The software structure and its actions are modified to allow for the detection and location of a fault, as well as preventing its propagation through the system. In this section, we describe common techniques for fault detection, fault containment, and fault recovery in software systems.

7.2.1 Fault Detection Techniques

As in the case of hardware, the goal of fault detection in software is to determine if a fault has occurred within a system. Single-version fault tolerance techniques usually use various types of *acceptance tests* to detect faults. The result of a program is subjected to a test. If the result passes the test, the program continues its execution. A failed test indicates a fault. A test is most efficient if it can be checked quickly. If a test takes nearly the same time as the program, it might be preferable to create a diverse version of the program and to compare the results of the two versions. Another desirable feature of acceptance tests is application independence. If a test can be developed based on criteria independent of the program application, it can be re-used in other applications.

The existing acceptance test techniques include timing checks, coding checks, reversal checks, reasonableness checks, and structural checks [27].

Timing checks are applicable to programs whose specification includes timing constraints. Based on these constraints, checks can be developed to indicate a deviation from the required behavior. A *watchdog timer* is an example of a timing check. Watchdog timers can be used to monitor the performance of a program and detect lost or locked out modules.

Coding checks are applicable to programs whose data can be encoded. For example, a cyclic code such as CRC can be used if the information is merely transported from one module to another without modifying its content. An arithmetic code can be used to detect errors in arithmetic operations.

In some programs, it is possible to reverse the output values and to compute the corresponding input values. In this case, *reversal checks* can be applied. A reversal check compares the actual inputs of a program with the computed ones. A disagreement indicates a fault.

Reasonableness checks use semantic properties of data to detect faults. For example, a range of data can be examined for overflow or underflow to indicate a deviation from the system's requirements.

Structural checks are based on known properties of data structures. For example, a number of elements in a list can be counted, or links and pointers can be verified. Structural checks can be made more efficient by adding redundant data to a data structure, e.g., attaching counts of the number of items in a list, or adding extra pointers [9].

7.2.2 Fault Containment Techniques

Fault containment in software can be achieved by modifying the structure of the system and by imposing a set of restrictions defining which actions are permissible within the system. Four common techniques for fault containment in software are: modularization, partitioning, system closure, and atomic actions.

Modularization attempts to prevent the propagation of faults by decomposing a system into modules, eliminating shared resources, and limiting the amount of communication between modules to carefully monitored messages [46]. Before performing the modularization, visibility, and connectivity parameters are examined to determine which module possesses the highest potential to cause the system failure [42]. The *visibility* of a module is characterized by the set of modules that may be invoked directly or indirectly by the module. The *connectivity* of a module is described by the set of modules that may be invoked directly or used by the module.

The isolation between functionally independent modules can be achieved by *partitioning*, the modular hierarchy of a software system in horizontal or vertical dimensions. The horizontal partitioning separates the major software functions into independent branches. The execution of the functions and the communication between the branches is done using control modules. Vertical partitioning distributes the control and processing function in a top-down hierarchy. High-level modules normally focus on control functions, while low-level modules perform processing [42].

Another technique used for fault containment in software is *system closure*. This technique is based on the principle that no action is permissible unless explicitly authorized [17]. Any component of a system is granted only the minimal capability which is required for performing its function. In an environment with many restrictions and strict control, e.g., in prison, all the interactions between the components of the system are visible. Therefore, it is easier to contain any fault.

An alternative technique for fault containment uses *atomic actions* to define interactions between system components. An atomic action among a group of components is an activity in which the components interact exclusively with each other [27]. There is no interaction with the rest of the system for the duration of the activity. Within an atomic action, the participating components neither import nor export any type of information from nonparticipating components of the system. An atomic action can have two possible outcomes: either it terminates normally, or it is aborted upon a fault detection. If an atomic action terminates normally, its results are correct. If a fault is detected, then this fault affects only the participating components. Thus, the fault containment area is well defined and fault recovery can be limited to the atomic action components only.

7.2.3 Fault Recovery Techniques

Once a fault is detected and contained, a system attempts to recover from the faulty state and regain operational status. If fault detection and containment mechanisms are implemented properly, the effects of the faults are contained within a particular set of modules at the moment of fault detection. Knowledge of the fault containment region is essential for the design of an effective fault recovery mechanism.

In this section, we consider two common techniques for fault recovery in software systems: checkpoint and restart and process pairs. We also describe a typical mechanism for initiation of fault recovery called exception handling.

7.2.3.1 Exception Handling

In many software systems, the request for initiation of fault recovery is issued by exception handling. *Exception handling* is the interruption of the normal operation to handle abnormal conditions [16]. Examples of abnormal conditions are out-of-range inputs values, memory corruption, null pointer, undefined state, and so on. If these conditions are not properly handled, they may cause the failure of a system. Failures due to abnormal conditions are estimated to cause two-thirds of system crashes and 50 % of system security problems [35]. Traditional software engineering techniques such as software testing [37] can catch and illuminate some abnormal conditions. However, it is impossible to cover all cases in a large software system. Exception handling is especially important in embedded systems in which software cannot easily be fixed or replaced [34].

Possible events triggering exceptions in a software module can be classified into three groups [43]:

1. *Interface exceptions* are signaled by a module when it detects an invalid service request. This type of exception is handled by the module that requested the service.
2. *Local exceptions* are signaled by a module when its fault-detection mechanism finds a fault within its internal operations. This type of exception is handled by the faulty module.
3. *Failure exceptions* are signaled by a module when it has detected that its fault recovery mechanism is unable to recover successfully. This type of exception is handled by the system.

7.2.3.2 Checkpoint and Restart

A popular recovery mechanism for single-version software fault tolerance is *checkpoint and restart* [15]. As mentioned previously, most of the software faults are design faults, activated by some unexpected input sequence. These types of faults resemble hardware intermittent faults: they appear for a short period of time, then disappear, and then appear again. Therefore, simply restarting the module is usually enough to successfully complete its execution [21].

The general scheme of the checkpoint and restart recovery mechanism is shown in Fig. 7.1. The module executing a program operates in combination with an acceptance test (AT) block which checks the correctness of the results. If a fault is detected, a "retry" signal is sent to the module to re-initialize its state to the checkpoint state stored in the checkpoint memory.

There are two types of checkpoints: static and dynamic [49]. A static checkpoint takes a single snapshot of the system state at the beginning of the program execution and stores it in the memory. Fault-detection checks are usually placed at the output of the module. If a fault is detected, the system returns to this state and starts the execution from the beginning. Dynamic checkpoints are created dynamically at various points during the execution. If a fault is detected, the system returns to the last checkpoint and continues the execution. Fault-detection checks need to be embedded in the code and executed before the checkpoints are created.

A number of factors influence the efficiency of checkpointing, including execution requirements, the interval between checkpoints, fault activation rate, and overhead associated with creating fault detection checks, checkpoints, recovery, etc [25]. In a static approach, the expected time to complete the execution grows exponentially with the processing requirements. Therefore, static checkpointing is effective only if the processing requirements are relatively small. In a dynamic approach, it is possible to achieve a linear increase in execution time as the processing requirements grow. There are three strategies for dynamic placing of checkpoints [39]:

1. *Equidistant*, which places checkpoints at deterministic fixed time intervals. The time between checkpoints is chosen depending on the expected fault rate.
2. *Modular*, which places checkpoints at the end of the submodules in a module, after the fault-detection checks for the submodule are completed. The execution time depends on the distribution of the submodules and expected fault rate.
3. *Random*, which places checkpoints at random.

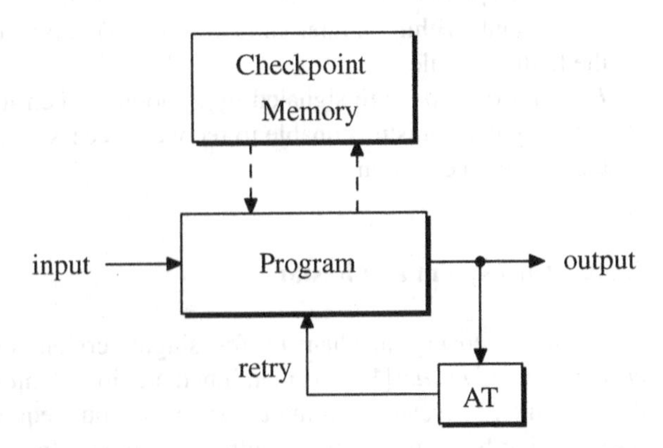

Fig. 7.1 Checkpoint and restart recovery

Example 7.1. A program is represented by the flowgraph shown in Fig. 7.2. The worst-case execution time of each block is shown inside the block. Decide where to place checkpoints if the process is required to return from the last checkpoint to the point where a fault occurred within 10 units of time. Assume that a block can be partitioned into subblocks. Minimize the number of checkpoints.

A possible solution is shown in Fig. 7.3. Four checkpoints are required.

Overall, the checkpoint and restart recovery mechanism has the following advantages [49]:

- It is conceptually simple.
- It is independent of the damage caused by a fault.
- It is applicable to unanticipated faults.
- It is general enough to be used at multiple levels in a system.

A problem with restart recovery is that *nonrecoverable actions* exist in some systems [27]. These actions are usually associated with external events that cannot be compensated by simply reloading the state and restarting the system. Examples of nonrecoverable actions are firing a missile or soldering a pair of wires [49]. The recovery from such actions needs to include special treatment, for example by compensating for their consequences or delaying their output until after additional confirmation checks are completed.

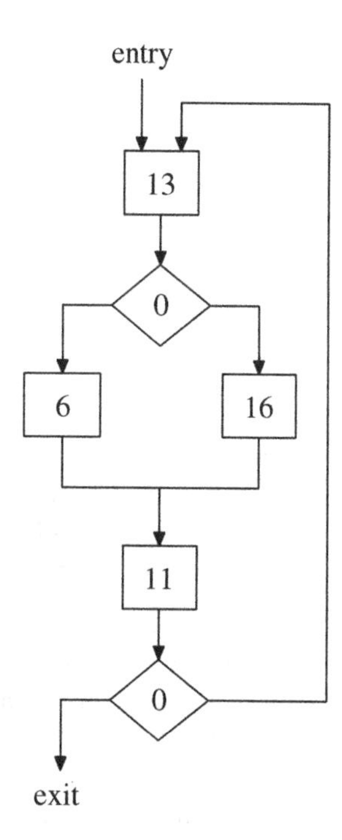

Fig. 7.2 Flowgraph for Example 7.1

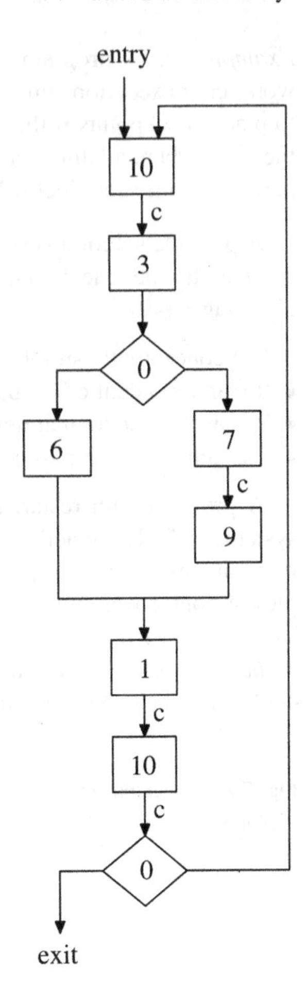

Fig. 7.3 Flowgraph with checkpoints for Example 7.1

Data Diversity

The efficiency of checkpoint and restart can be improved by using different input re-expressions for each retry. Since software faults are usually input sequence dependent, if inputs are re-expressed in a diverse way, it is unlikely that different re-expressions activate the same fault. Such a technique is called *data diversity* [1].

There are three basic techniques for data diversity [27]:

1. Input data re-expression in which only the input is changed and no postexecution adjustment is required.
2. Input data re-expression with postexecution adjustment. In this case, the output result has to be adjusted in accordance with a given set of rules. For example, if the inputs were re-expressed by encoding them in some code, then the output result is decoded following the decoding rules of the code.

3. Input data re-expression via decomposition and re-combination. In this case, the input is decomposed into smaller parts and then re-combined to obtain the output result.

Example 7.2. Suppose that a program code contains an expression of a Boolean function, e.g.

$$f(a, b, c) = ab + ac.$$

Then, a data-diverse version of this code may contain a functionally equivalent, but different expression of the same Boolean function, e.g.,

$$f(a, b, c) = a(b + c).$$

This is an example of input data re-expression in which no postexecution adjustment is required.

Alternatively, we can apply decomposition and re-combination to compute $f(a, b)$ as:

$$g(b, c) = b + c$$
$$f(a, b, c) = a \cdot g(b, c).$$

7.2.3.3 Process Pairs

Process pair technique [18] runs two identical versions of the software on separate processors. The block diagrams are shown in Fig. 7.4. First the primary processor, Processor 1, is active. It executes the program and sends the checkpoint information to the secondary processor, Processor 2. If a fault is detected by AT block, the primary processor is switched off. The secondary processor loads the last checkpoint as its starting state and continues the execution. Meanwhile, Processor 1 executes diagnostic checks off-line. If the fault is nonrecoverable, the replacement is performed. After returning to service, the repaired processor becomes the secondary processor.

Process pairs are used, for example, to access I/O devices in Tandem computers [6]. One I/O process is designated as active, while the other one serves as a backup. All messages are delivered through the primary I/O process. If the primary process fails, the secondary one takes over.

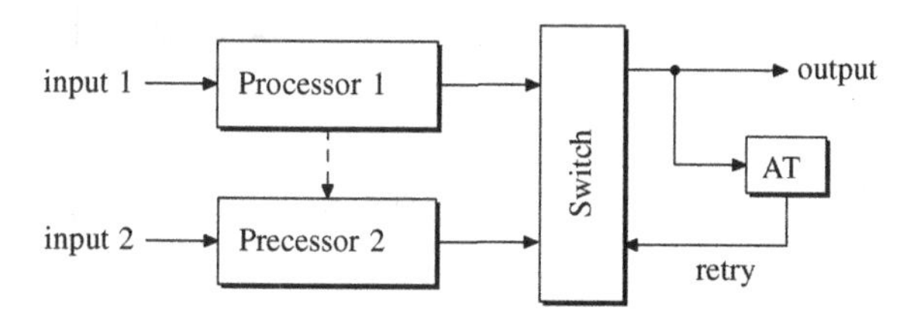

Fig. 7.4 Process pairs

The main advantage of the process pair technique is that it continues providing uninterrupted service after the occurrence of the fault. It is therefore suitable for applications requiring high availability [49].

7.3 Multi-Version Techniques

Multi-version techniques use multiple versions of the same software component, which are developed following design diversity rules. For example, different teams, different coding languages, or different algorithms can be used to maximize the probability that different versions do not have common faults. In this section, we describe three popular multi-version techniques: recovery blocks, N-version programming, and N self-checking programming.

7.3.1 Recovery Blocks

The recovery block technique applies checkpoint and restart to multiple versions of a software module. Its configuration is shown in Fig. 7.5. Versions 1 to n represent diverse but functionally equivalent versions of the same module. In the beginning, Version 1 provides the system's output. If an error is detected by the acceptance test, a retry signal is sent to the switch. The system is rolled back to the last state stored in the checkpoint memory and the execution is switched to the next version. Checkpoints are created every time before a new version executes. If all n versions are tried and none of the results are accepted, the system fails.

Various checks are used for acceptance testing of the active version of the module. Checks are placed either at the output of a module, or embedded in the code to increase the effectiveness of fault detection.

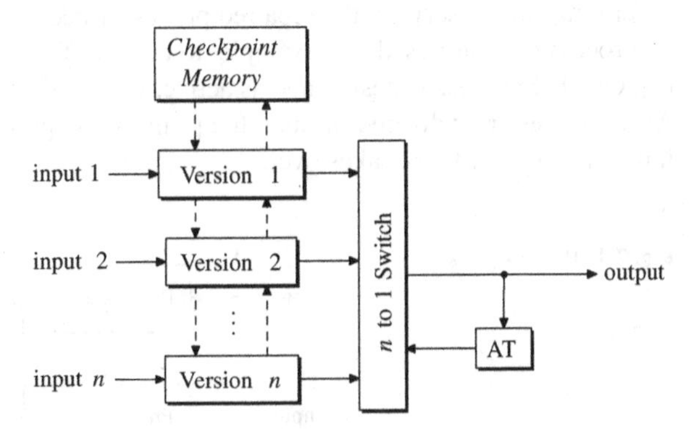

Fig. 7.5 Recovery block technique

Table 7.1 States of a Markov chain for Example 7.3

State	Description
1	Three versions and switch are operational
2	One version failed, two versions and switch are operational
3	Two versions failed, one version and switch are operational
4	The system failed

Like all multi-version techniques, the recovery block technique is heavily dependent on design diversity. It requires that the specification of the module is detailed enough to allow for the creation of multiple alternatives that are functionally equivalent. This issue is further discussed in Sect. 7.3.4. Another problem is that, usually, acceptance tests are highly application dependent and difficult to create. They cannot test for a specific answer, but only for "acceptable" values [44].

Example 7.3. If we expect a recovery block system to be diagnosed and restarted within a short time interval, then we can assume constant failure and restart rates [2]. With this assumption, Markov chains can be used for modeling the system.

Draw a Markov chain for reliability analysis of the recovery block system composed of three diverse versions of the same module. Assume that:

- the failures and repairs of different versions are independent events,
- the failure rate of each version is λ and the restart rate is ρ,
- the failure rate of the switch is λ_s,
- the checkpoint memory and acceptance test block and switch cannot fail.

The resulting Markov chain is shown in Fig. 7.6. The states are labeled according to Table 7.1. States 1–3 are operational states. State 4 is failed state. In a recovery block system, it is sufficient that one version is operational for the system to be operational. Since we perform a reliability analysis, the system cannot be repaired from a failed state. Therefore, we do not need to distinguish between failures of a system due to the failure of the switch or to the failure of a version. All failed states can be merged into one.

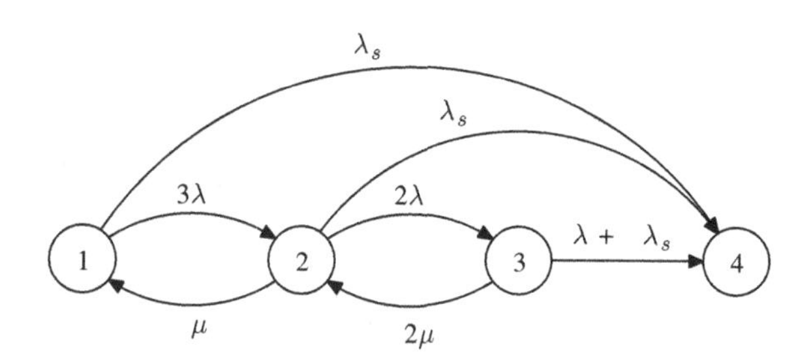

Fig. 7.6 Markov chain for Example 7.3

7.3.2 N-Version Programming

The N-version programming technique [5] resembles the N-modular hardware redundancy. The block diagram is shown in Fig. 7.7. It consists of n diverse software implementations of the same module, executed concurrently. All versions are functionally equivalent. The selection algorithm decides which of the answers is correct and returns this answer as a result of the program execution. The selection algorithm is usually implemented as a generic voter. This is an advantage over the recovery blocks fault-detection mechanism, which requires an application-dependent acceptance test.

Many different types of voters for software systems have been developed, including the formalized majority voter, the generalized median voter, the formalized plurality voter, and the weighted averaging technique [31]. These voters have the capability of performing inexact voting by using the concept of *metric space* (X, d). The set X is the output space of the software and d is a metric function that associates any two elements in X with a real-valued number (see Sect. 5.2.5 for the definition of metric). The inexact values are declared equal if their metric distance is less than some pre-defined threshold ε.

In the *formalized majority voter*, the outputs are compared and, if more than half of the values agree, the voter output is selected as one of the values in the agreement group.

The *generalized median voter* selects the median of the values as the correct result. The median is computed by successively eliminating pairs of values that are farther apart until only one value remains. The number of outputs is required to be odd.

The *formalized plurality voter* partitions the set of outputs based on metric equality and selects the output from the largest partition group.

The *weighted averaging technique* combines the outputs in a weighted average to produce the result. The weight can be selected in advance based on the characteristics of the individual versions. If all the weights are equal, this technique reduces to the median selection technique which we considered in Sect. 4.2.1.2. The weight can be also selected dynamically based on pair-wise distances of the version outputs [20] or the success history of the versions measured by some performance metric [11].

The selection algorithms are normally developed taking into account the consequences of erroneous output for the system. For applications where reliability is

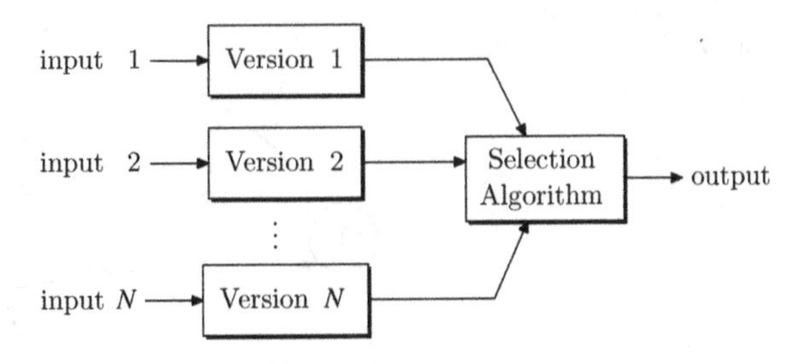

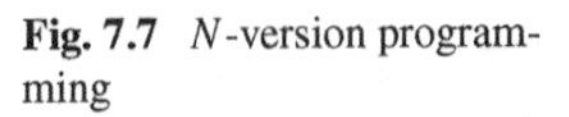
Fig. 7.7 N-version programming

important, the selection algorithms are designed, so that the selected result is correct with a very high probability. If availability is an issue, the selection algorithm may occasionally produce an output even if it is incorrect. Such an outcome is acceptable as long as the program execution is not subsequently dependent on previously generated, possibly erroneous results. For applications where safety is the main concern, the selection algorithm is required to correctly distinguish the erroneous result and mask it. In cases when the algorithm cannot select the correct result with a high confidence, it should report the error to the system or initiate an acceptable safe output sequence [49].

N-version programming technique can tolerate design faults presented in software if the design diversity is implemented properly. Each version of the module should be implemented by using different tool sets, different programming languages, and possibly different environments. The various development groups must have as little interaction as possible. The specification should be flexible enough and detailed enough to give the programmer a possibility to create diverse but compatible versions.

There are a number of differences between the recovery block technique and the N-version programming. In traditional recovery blocks, each alternative version is executed serially until a fault-free solution is found by the checker. In the N-version programming technique, versions are normally run concurrently. On the one hand, the time to serially try multiple alternatives in recovery blocks may be prohibitive, especially for real-time systems. On the other hand, running N-versions concurrently requires N redundant hardware modules.

Another difference is the mechanism for fault detection. The recovery block technique requires an application-dependent acceptance test and individual checks for each of the versions, which are either embedded in the code, or placed at the output of the module. In N-version programming, a generic voter can be used. Since either of the techniques has its own pros and cons, a careful analysis needs to be performed to determine which of them is best for a given system.

Example 7.4. A 3-version programming system consists of three diverse software implementations of the same module. The probabilities that the versions 1, 2, and 3 are faulty are 0.1, 0.12, and 0.15 per 4×10^4 s of CPU execution time, respectively. Assuming that the versions are truly diverse and their failures are independent events, what is the reliability of the resulting system per 4×10^4 s of CPU execution time given that the reliability of the voter is 0.94 per 4×10^4 s of CPU execution time?

If we assume that failures of different versions are independent events, then the expression for reliability of the TMR can be used to evaluate the reliability of the 3-version programming:

$$R_{3-\text{version}} = (R_1 R_2 R_3 + (1 - R_1) R_2 R_3 + R_1 (1 - R_2) R_3 + R_1 R_2 (1 - R_3)) R_v.$$

By substituting $R_1 = 0.9$, $R_2 = 0.88$, $R_3 = 0.85$, and $R_v = 0.94$, we get

$$R_{3-\text{version}}(4 \times 10^4 \text{ s}) = 0.901.$$

7.3.3 N Self-Checking Programming

N self-checking programming combines recovery blocks and N version programming [26]. The checking is performed either by using acceptance tests, or by comparing pairs of modules. Examples of applications of N self-checking programming are the 5ESS Switching System which services approximately half of all US telephone exchanges [13] and the Airbus A-340 flight control computer [10].

The structure of an N self-checking programming system using acceptance tests is shown in Fig. 7.8. Different versions of a software module and the acceptance tests are developed independently from a common specification. The individual checks for each of the versions are either embedded in the code, or placed at the output of the modules. The use of separate acceptance tests for each version is the main difference from the recovery block approach. The execution of each version can be done either serially, or concurrently. In both cases, the output is taken from the highest-ranking version which passes its acceptance test.

The structure of an N self-checking programming system using comparison is shown in Fig. 7.9. An advantage over the N self-checking programming using acceptance tests is that an application-independent decision algorithm is used for fault detection.

Example 7.5. Derive an expression for the reliability of the five self-checking programming system using acceptance tests in which different versions are executed concurrently. Assume that failures of different versions are independent events, the version i has the reliability R_i, for $i \in \{1, 2, 3, 4, 5\}$, the reliability of the switch is R_s, and the acceptance test blocks are perfect.

In the five self-checking programming system, it is sufficient that one version is operational for the system to be operational. Therefore, its reliability can be computed as 1 minus the probability that all five modules failed multiplied by the reliability of the switch:

$$R_{5sc} = (1 - (1 - R_1)(1 - R_2)(1 - R_3)(1 - R_4)(1 - R_5))R_s.$$

Fig. 7.8 N self-checking programming using acceptance tests

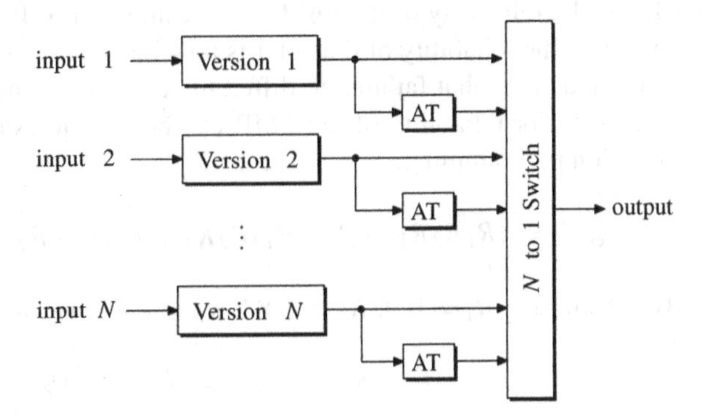

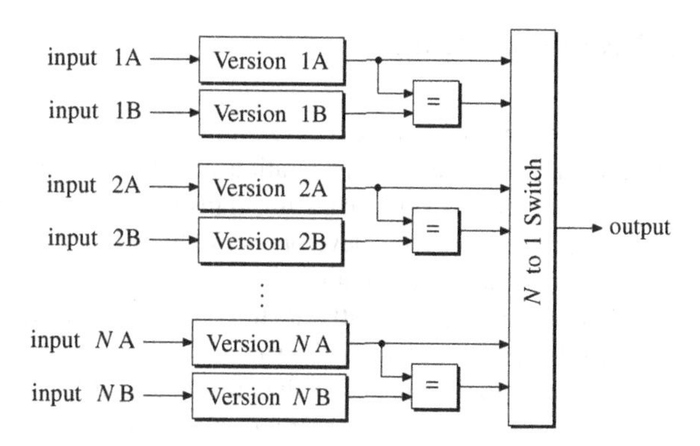

Fig. 7.9 *N* self-checking programming using comparison

7.3.4 Importance of Design Diversity

The most critical issue in multi-version techniques is assuring independence between the different versions of a software module through design diversity. Design diversity aims to protect multiple versions of a module from common design faults [4]. Software systems are vulnerable to common design faults if they implement the same algorithm, use the same program language, or if they are developed by the same design team, tested using the same technique, etc.

Presently, the implementation of design diversity remains a controversial subject [24]. The increase in complexity caused by redundant multiple versions can be quite severe and may result in a less dependent system, unless appropriate measures are taken. Decisions to be made when developing a multi-version software system include [5]:

- which modules are to be made redundant;
- the level of redundancy (procedure, process, and whole system);
- the required number of redundant versions;
- the required diversity (diverse specification, algorithm, code, programming language, team, testing technique, and so on.);
- rules of isolation between the development teams, to prevent the flow of information that could result in common design error.

The cost of developing multi-version software also needs to be taken into account. A direct replication of the full development effort is prohibitive for most applications. The cost can be reduced by allocating redundancy to the dependability-critical parts of a system only [8].

7.4 Software Testing

Software testing is the process of executing a program with the intent of finding errors [37]. Testing is a major consideration in software development. In many organizations, more time is devoted to testing than to any other phase of software development. On complex projects, test developers might be twice or three times as many as code developers in the project team.

There are two types of software testing: functional and structural [45]. *Functional testing* (also called *behavioral testing*, *black-box testing*, *closed-box testing*), compares test program behavior against its specification. *Structural testing* (also called *white-box testing*, *glass-box testing*) checks the internal structure of a program for errors. For example, suppose we test a program which adds two integers. The goal of functional testing is to verify whether the implemented operation is indeed addition instead of, for example, multiplication. Structural testing does not question the functionally of the program, but checks whether the internal structure is consistent. One strength of the structural approach is that the entire software implementation is taken into account during testing, which facilitates error detection even if the software specification is vague or incomplete.

The effectiveness of structural testing is normally expressed in terms of test coverage metrics which measure the fraction of code exercised by test cases. Common test coverage metrics are statement, branch, and path coverage [7]. *Statement* coverage requires that the program under test is run with enough test cases, so that all its statements are executed at least once. *Decision* coverage requires that all branches of the program are executed at least once. *Path* coverage requires that each of the possible paths through the program is followed. Path coverage is the most reliable metric; however, it is not applicable to large systems, since the number of paths is exponential to the number of branches.

This section gives a brief overview of statement and branch coverage metrics.

7.4.1 Statement Coverage

Statement coverage (also called *line coverage* or, *segment coverage* [40]) examines whether each executable statement of a program is followed during a test. An extension of statement coverage is *basic block coverage* [48], in which each sequence of nonbranching statements is treated as one statement unit.

The main advantage of statement coverage is that it can be applied directly to object code and does not require processing source code. Its disadvantages are:

- Statement coverage is insensitive to some control structures, logical AND and OR operators, and switch labels.
- Statement coverage only checks whether the loop body was executed or not. It does not report whether loops reach their termination condition. In C, C++, and Java programs, this limitation affects loops that contain break statements.

As an example of the insensitivity of statement coverage to some control structures, consider the following code:

```
x = 0;
if (condition)
       x = x + 1;
y = 10/x
```

If there is no test case which causes the `condition` to evaluate false, the error in this code will not be detected in spite of 100 % statement coverage. The error will appear only if the `condition` evaluates false for some test case. Since `if`-statements are common in programs, this problem is a serious drawback of statement coverage.

7.4.2 Branch Coverage

Branch coverage (also referred to as *decision coverage* or *all-edges coverage* [40]) requires that each branch of a program is executed at least once during a test. Boolean expressions of `if`- or `while`-statements are checked to be evaluated both true and false. The entire Boolean expression is treated as one predicate regardless of whether it contains logical AND and OR operators. `Switch` statements, exception handlers, and interrupt handlers are treated similarly. Decision coverage includes statement coverage since executing every branch leads to executing every statement.

Branch coverage allows many problems of statement coverage to be overcome. However, it might miss some errors, as demonstrated by the following example:

```
if (condition1)
       x = 0;
else
       x = 2;
if (condition2)
       y = 10*x;
else
       y = 10/x;
```

A 100 % branch coverage can be achieved by two test cases which cause both `condition1` and `condition2` to evaluate true, and both `condition1` and `condition2` to evaluate false. However, the error which occurs when `condition1` evaluates true and `condition2` evaluates false will not be detected by these two tests.

The error in the example above can be detected by exercising every path through the program. However, since the number of paths is exponential to the number of branches, testing every path is not possible for large systems. For example, if one test case takes 0.1×10^{-5} s to execute, then testing all paths of a program containing 30 `if`-statements will take 18 min and testing all paths of a program with 60 `if`-statements will take 366 centuries [19].

7.5 Summary

In this chapter, we have considered how fault tolerance can be implemented in the software domain. We have studied the most popular single-version and multi-version software fault tolerance techniques. We have discussed the importance of design diversity and software testing. We have analyzed common test coverage metrics.

Problems

7.1 Give three examples of software faults, software errors, and software failures.

7.2 Some hardware faults have a similar nature to software faults. Give three examples of such faults.

7.3 A program is represented by the flowgraph shown in Fig. 7.2. The worst-case execution time of each block is shown inside the block. Decide where to place checkpoints if the process is required to return from the last checkpoint to the point where a failure occurred within 12 units of time. Assume that a block can be partitioned into subblocks. Minimize the number of checkpoints.

7.4 The execution time of a program is T. Suppose that n checkpoints have been inserted into the program at equal time intervals. The time overhead of each checkpoint is t. What is the worst-case execution time of the program if k failures are expected? Assume that the time overhead of each checkpoint is t and the fault recovery time is negligibly small.

7.5 The following software reliability growth model, taken from [47], can be used to estimate the failure rate of initially released software and the improved failure rate which can be expected after debugging. The initial software failure rate is computed as:

$$\lambda_0 = \frac{r_i K W_0}{n} \quad \text{failures per CPU second}$$

where r_i is the host processor speed expressed in instructions per second, K is the fault exposure ratio, which is a function of program structure and data dependency, W_0 is the estimate of the total number of faults in the initial program, and n in the number of object instructions given by the number of source code lines times the expansion ratio.

The software failure rate at time t is computed as:

$$\lambda(t) = \lambda_0 e^{-\beta t}$$

where t is CPU time and $\beta = B\frac{\lambda_0}{W_0}$ is the decrease in failure rate per failure occurrence, where B is the fault reduction factor.

Estimate the initial failure rate and the failure rate after 5×10^4 s of CPU execution time for 3×10^4 line C program. Use the following parameters:

- $r_i = 2 \times 10^7$ instructions per second
- $K = 4.2 \times 10^7$
- $W_0 = 6$ faults per 1,000 lines of code
- expansion ratio = 2.5
- $B = 0.955$

7.6 An engineer designs a software system consisting of three modules in series with the reliabilities $R_1 = 0.89$, $R_2 = 0.85$, and $R_2 = 0.83$ per 6×10^4 s of CPU execution time. It is acceptable to add two redundant modules to the system (diverse versions). Which of the following is best to do:

1. Duplicate modules 2 and 3 using high-level redundancy (as in Fig. 4.2a).
2. Duplicate modules 2 and 3 using low-level redundancy (as in Fig. 4.2b).
3. Triplicate the module 3.

7.7 Draw a Markov chain for availability analysis of the recovery block system composed on three diverse versions of the same module. Assume that:

- the failure rate of each version is λ and the restart rate is ρ,
- when the first version fails, the failure rates of the remaining versions increase to λ'. When the second version fails, the failure rates of the third version become λ'',
- repairs of different versions are independent events,
- the failure rate of the switch is λ_s and the restart rate is ρ_s,
- the checkpoint memory, acceptance test block and switch cannot fail.

7.8 Draw a Markov chain for safety analysis of the recovery block system composed of three diverse versions of the same module. Assume that:

- the failure rate of each version is λ and the restart rate is ρ,
- failures and repairs of different versions are independent events,
- the probability that a faulty module passes the acceptance test is q,
- the checkpoint memory, acceptance test block, and switch cannot fail,
- the system cannot be restarted if it failed nonsafe.

7.9 A program consists of seven independent modules which are all required for the normal operation of the program. The probability that each module is faulty is 0.12 per 6×10^4 s of CPU execution time. It is planned to use 3-version programming with voting after the execution of each module. The reliability of the voter is 0.9 per 6×10^4 s of CPU execution time. Assuming that module failures are independent events, what is the reliability of the program per 6×10^4 s of CPU execution time:

1. When a nonredundant version is used and no testing is performed.
2. When a nonredundant version is used, but extensive testing is performed that reduces the fault content of each module to 15 % of the original level.
3. When 3-version programming is used and no testing is performed? Assume that the versions are truly diverse and their failures are independent events.

7.10 Repeat Problem 7.9 (3) for the case when a single voter at the end of the execution of all seven modules is used.

7.11 N diverse software versions of a module are developed, each having a failure probability of 0.1 per 30,000 s of CPU execution time. Assuming that failures of different versions are independent events and the voter is perfect, which value of N should be used for an N-version programming system to have the failure probability of no more than 3 % per 30,000 s of CPU execution time?

7.12 Derive an expression for the reliability of the N self-checking programming using comparison composed of six diverse versions which are executed concurrently. Assume that failures of the different versions are independent events, all versions have the same reliability R, the acceptance test blocks are perfect, and the reliability of the switch is R_s.

7.13 How many module faults can you tolerate in N self-checking programming using acceptance tests?

7.14 How many module faults can you tolerate in N self-checking programming using comparison?

7.15 Give an example of a code in which a bug will not be detected in spite of 100 % statement coverage. Will this bug be detected by 100 % branch coverage?

7.16 Give an example of a code in which a bug will not be detected in spite of 100 % branch coverage.

7.17 A *kernel* K of a flowgraph F is a subset of vertices of F which satisfies the property that any set of tests which executes all vertices of the kernel executes all vertices of F [19]. 100 % statement coverage can be achieved by constructing test cases for vertices of a kernel only. Minimal kernels can be computed using dominator relations of a flowgraph.

A vertex v *pre-dominates* another vertex u, if every path from $entry$ to u contains v. A vertex v *post-dominates* another vertex u, if every path from u to $exit$ contains v. Vertex v is the *immediate dominator* of u, denoted by $idom(v)$, if v dominates u and every other dominator of u dominates v. The edges $(idom(v), v)$ form a directed tree rooted at $entry$ for pre-dominators and at $exit$ for post-dominators [32].

A minimum kernel set for G can be computed as $K = L - P$ where L is the set of leaf vertices of the post-dominator tree of G and P is a subset of vertices of L which predominate some vertex of L [19].

Compute a minimum kernel set for the flowgraph in Fig. 7.10.

7.18 The kernel-based technique described in Problem 7.17 can be similarly applied to branch coverage by constructing pre-and post-dominator trees for the edges of the flowgraph instead of for its vertices. A kernel set for edges is defined as a subset of edges of the flowgraph which satisfies the property that any set of tests which executes all edges of the kernel executes all edges of the flowgraph. 100 % branch coverage can be achieved by constructing test cases for edges of a kernel only.

Compute a minimum kernel set for the edges of the flowgraph in Fig. 7.10.

Fig. 7.10 Flowgraph for Problem 7.17

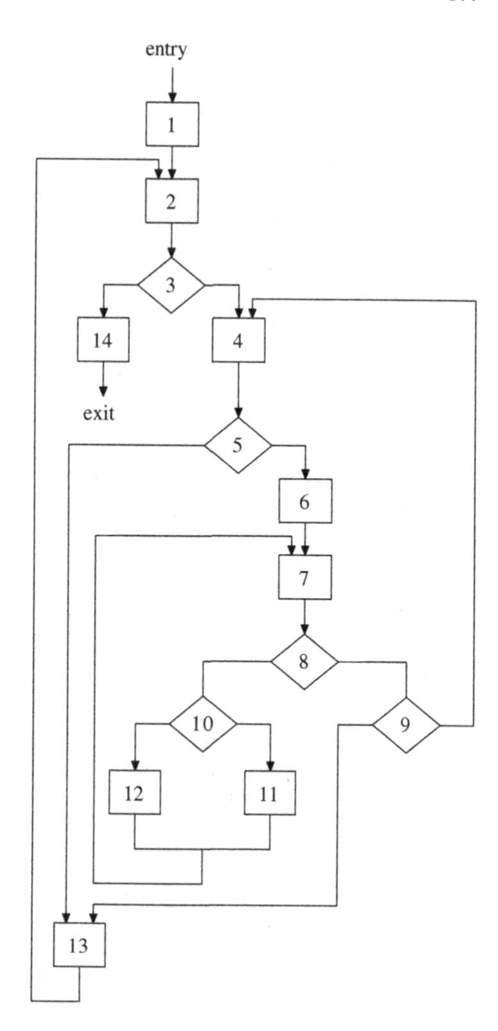

References

1. Ammann, P., Knight, J.: Data diversity: an approach to software fault tolerance. IEEE Trans. Comput. **37**(4), 418–425 (1988)
2. Aveyard, R.L., Man, F.T.: A study on the reliability of the circuit maintenance system-1 b. Bell Syst. Tech. J. **59**(8), 1317–1332 (1980)
3. Avižienis, A.: Fault-tolerant systems. IEEE Trans. Comput. **25**(12), 1304–1312 (1976)
4. Avižienis, A.: Design diversity: an approach to fault tolerance of design faults. In: Proceedings of the National Computer Conference and Exposition, pp. 163–171 (1984)
5. Avižienis, A.: The methodology of N-version programming. In: Lyu, M.R. (ed.) Software Fault Tolerance. Wiley, Chichester, pp. 158–168 (1995)
6. Bartlett, J.F.: A "NonStop" operating system. In: Proceedings of the 11th Hawaii International Conference on System Sciences, vol. 3 (1978)
7. Beizer, B.: Software Testing Techniques. Van Nostrand Reinhold, New York (1990)

8. Bishop, P.: Software fault tolerance by design diversity. In: Lyu, M.R. (ed.) Software Fault Tolerance. Wiley, New York, pp. 211–229 (1995)
9. Black, J.P., Taylor, D.J., Morgan, D.E.: An introduction to robust data structures. Computer Science Department, University of Waterloo, CS-80-08. Computer Science Department, University of Waterloo (1980)
10. Briere, D., Traverse, P.: AIRBUS A320/A330/A340 electrical flight controls—a family of fault-tolerant systems. In: Digest of Papers of The Twenty-Third International Symposium on Fault-Tolerant, Computing (FTCS'93), pp. 616–623 (1993)
11. Broen, R.B.: New voters for redundant systems. J. Dyn. Syst. Meas. Control **97**(1), 41–45 (1975)
12. Brooks, F.P.: No silver bullet: essence and accidents of software engineering. IEEE Comput. **20**(4), 10–19 (1987)
13. Carney, D., Cochrane, J.: The 5ESS switching system: architectural overview. AT&T Tech. J. **64**(6), 1339–1356 (1985)
14. Challenger: report of on the space shuttle Challenger accident. http://science.ksc.nasa.gov/shuttle/missions/51-l/docs/rogers-commission/tab (1986)
15. Chandy, K., Browne, J., Dissly, C., Uhrig, W.: Analytic models for rollback and recovery strategies in data base systems. IEEE Trans. Softw. Eng. **SE-1**(1), 100–110 (1975)
16. Cristian, F.: Exception handling and software fault tolerance. IEEE Trans. Comput. **C-31**(6), 531–540 (1982)
17. Denning, P.J.: Fault tolerant operating systems. ACM Comput. Surv. **8**(4), 359–389 (1976)
18. Dimmer, C.I.: The tandem non-stop system. In: Anderson, T. (ed.) Resilient Computing Systems, vol. 1. Wiley, New York, pp. 178–196 (1986)
19. Dubrova, E.: Structural testing based on minimum kernels. In: Proceedings of the Conference on Design, Automation and Test in Europe—Volume 2, DATE '05, pp. 1168–1173 (2005)
20. Gersting, J., Nist, R., Roberts, D., Van Valkenburg, R.: A comparison of voting algorithms for N-version programming. In: Proceedings of the Twenty-Fourth Annual Hawaii International Conference on System Sciences, vol. ii, pp. 253–262 (1991)
21. Gray, J.: Why do computers stop and what can be done about it? In: Proceedings of the the Fifth Symposium of Reliability in Distributed Software and Database Systems, pp. 3–12 (1986)
22. Hailpern, B., Santhanam, P.: Software debugging, testing, and verification. IBM Syst. J. **41**(1), 4–12 (2002)
23. Heimdahl, M.P.E., Leveson, N.G.: Completeness and consistency in hierarchical state-based requirements. IEEE Trans. Softw. Eng. **22**(6), 363–377 (1996)
24. Koopman, P.: Better Embedded System Software. Drumnadrochit Press, Wilmington (2010)
25. Kulkarni, G.V., Nicola, F.V., Trivedi, S.K.: Effects of checkpointing and queueing on program performance. Commun. Stat. Stochast. Models **6**(4) 615–648 (1990)
26. Laprie, J.C., Arlat, J., Beounes, C., Kanoun, K.: Definition and analysis of hardware- and software-fault-tolerant architectures. Computer **23**(7), 39–51 (1990)
27. Lee, P.A., Anderson, T.: Fault tolerance: principles and Practice. Dependable computing and fault-tolerant systems, 2nd edn. Springer-Verlag, New York (1990)
28. Leveson, N., Turner, C.S.: An investigation of the Therac-25 accidents. IEEE Comput. **26**, 18–41 (1993)
29. Lin, H.: Sheffield hickups caused by software. Sci. Am. **253**(6), 48 (1985)
30. Lions, J.L.: Ariane 5 flight 501 failure, report by the inquiry board. http://www.esrin.esa.it/htdocs/tidc/Press/Press96/ariane5rep.html (1996)
31. Lorczak, P., Caglayan, A., Eckhardt, D.: A theoretical investigation of generalized voters for redundant systems. In: Nineteenth International Symposium on Fault-Tolerant Computing, 1989, FTCS-19. Digest of Papers, pp. 444–451 (1989)
32. Lowry, E.S., Medlock, C.W.: Object code optimization. Commun. ACM **12**(1), 13–22 (1969)
33. Lyu, M.R.: Introduction. In: Lyu, M.R. (ed.) Handbook of Software Reliability. McGraw-Hill, New York, pp. 3–25 (1996)
34. Massa, A.J.: Embedded development: Handling exceptions and interrupts in eCos. http://www.informit.com/articles/article.aspx?p=32058 (2003)

35. Maxion, R.A., Olszewski, R.T.: Improving software robustness with dependability cases. In: Proceedings of the The Twenty-Eighth Annual International Symposium on Fault-Tolerant Computing, pp. 346–355 (1998)
36. McAllister, D., Vouk, M.A.: Fault-tolerant software reliability engineering. In: Lyu, M.R. (ed.) Handbook of Software Reliability. McGraw-Hill, New York, pp. 567–614 (1996)
37. Myers, G.J.: Art of Software Testing. Wiley, New York (1979)
38. Myers, W.: Can software for the strategic defense initiative ever be error free? IEEE Comput. **19**(11), 61–67 (1986)
39. Nicola, V.F.: Checkpointing and the modeling of program execution time. In: Lyu, M.R. (ed.) Software Fault Tolerance. Wiley, Chichester, pp. 167–188 (1995)
40. Ntafos, S.: A comparison of some structural testing strategies. IEEE Trans. Softw. Eng. **14**(6), 868–874 (1988)
41. Pratt, V.: Anatomy of the Pentium bug. In: Mosses, P.D., Nielsen, M., Schwartzbach, M.I.(eds.) TAPSOFT'95: Theory and Practice of Software Development, vol. 915. Springer Verlag, Berlin, pp. 97–107 (1995)
42. Pressman, R.S.: Software Engineering: A Practitioner's Approach. The McGraw-Hill Companies, Inc., New York (1997)
43. Randell, B.: System structure for software fault tolerance. In: Proceedings of the International Conference on Reliable Software, pp. 437–449 (1975)
44. Randell, B., Xu, J.: The evolution of the recovery block concept. In: Lyu, M.R. (ed.) Software Fault Tolerance. Wiley, New York, pp. 1–21 (1995)
45. Roper, M.: Software Testing. McGraw-Hill Book Company, London (1994)
46. Rushby, J.M.: Bus architectures for safety-critical embedded systems. In: Proceedings of the First International Workshop on Embedded Software, EMSOFT '01, pp. 306–323 (2001)
47. TR-528-96: Reliability techniques for combined hardware and software systems. Technical Report TR-528-96, Rome Laboratory (1992)
48. Watson, A.H.: Structured testing: analysis and extensions. Technical Report TR-528-96, Princeton University (1996)
49. Wilfredo, T.: Software fault tolerance: a tutorial. Technical Report, Langley Research Center, Hampton (2000)
50. Woodcock, J., Larsen, P.G., Bicarregui, J., Fitzgerald, J.: Formal methods: practice and experience. ACM Comput. Surv. **41**(4), 19:1–19:36 (2009)

Chapter 8
Conclusion

The chapters in this book have covered the main concepts of fault tolerance, basic techniques for designing fault-tolerant hardware and software systems, and common methods for modeling and evaluating fault-tolerant systems in terms of reliability, availability, and safety.

There are certainly many important aspects of fault-tolerant design beyond the ones we have covered in this book, including specification, validation, and testing of fault-tolerant systems, dependability assessment using test data, failure mode and effects analysis, and security.

Fault tolerance is a fascinating and constantly developing field. We hope that this book has sparked your interest to explore it further. We wish you the best on your continued journey of learning.

E. Dubrova, *Fault-Tolerant Design*, DOI: 10.1007/978-1-4614-2113-9_8,

Index

E. Dubrova, *Fault-Tolerant Design*, DOI: 10.1007/978-1-4614-2113-9,